Palgrave Studies in Education and the Environment

This series focuses on new developments in the study of education and environment. Promoting theoretically-rich works, contributions include empirical and conceptual studies that advance critical analysis in environmental education and related fields. Concerned with the underlying assumptions and limitations of current educational theories in conceptualizing environmental and sustainability education, the series highlights works of theoretical depth and sophistication, accessibility and applicability, with critical orientations to matters of public concern. It engages interdisciplinary and diverse perspectives as these relate to domains of policy, practice, and research. Studies in the series may span a range of scales from the more micro level of empirical thick description to macro conceptual analyses, highlighting current and upcoming turns in theoretical thought. Tapping into a growing body of theoretical scholarship in this domain, the series provides a venue for examining and expanding theorizations and approaches to the interdisciplinary intersections of environment and education. Its timeliness is clear as education becomes a key mode of response to environmental and sustainability issues internationally. The series will offer fresh perspectives on a range of topics such as:

- curricular responses to contemporary accounts of human-environment relations (e.g., the Anthropocene, nature-culture, animal studies, transdisciplinary studies)
- the power and limits of new materialist perspectives for philosophies of education
- denial and other responses to climate change in education practice and theory
- place-based and land-based orientations to education and scholarship
- postcolonial and intersectional critiques of environmental education and its research
- policy research, horizons, and contexts in environmental and sustainability education

More information about this series at
http://www.palgrave.com/gp/series/15084

Stuart Tannock

Educating for Radical Social Transformation in the Climate Crisis

palgrave
macmillan

Stuart Tannock
UCL Institute of Education
LONDON, UK

ISSN 2662-6519 ISSN 2662-6527 (electronic)
Palgrave Studies in Education and the Environment
ISBN 978-3-030-82999-5 ISBN 978-3-030-83000-7 (eBook)
https://doi.org/10.1007/978-3-030-83000-7

Cover illustration: © frans lemmens / Alamy

This Palgrave Macmillan imprint is published by the registered company Springer Nature Switzerland AG.
The registered company address is: Gewerbestrasse 11, 6330 Cham, Switzerland

Acknowledgement

This book was written during an extraordinary period of social and intellectual isolation due to the Covid-19 health crisis that swept the world in 2020–2021. When I initially conceptualized this project, I envisioned a research process that would link theoretical and historical research with a series of empirical case studies of innovative examples of climate change education. Plague and lockdown confined me to my desk, and as it transpired, the theoretical and historical work provided more than enough material to complete a book: empirical studies would have made the book overly long and hopefully can be picked up in follow up work in the future.

However, while the writing was done in a state of relative isolation, the ideas in this book were developed through opportunities spanning over the last two decades to engage with and learn from many others. My first academic job was at the University of California, Berkeley, where I had the privilege of working with two different centers that have shaped my thinking about education, social justice, and social change ever since. One was the student-developed Center for Popular Education and Participatory Research (CPEPR), that in the early 2000s ran a series of workshops and conferences bringing together people from across North America working with traditions developed by Paulo Freire and others. Though I had read Freire previously, this was where I learned to think carefully about this vital tradition. I am deeply grateful to those involved in creating and running CPEPR, including: Andrea Dyrness, Emma Fuentes, Shabnam Koirala, Soo Ah Kwon, Kysa Ngreen, Patricia Sánchez, and the late Professor John Hurst.

I also had the opportunity at Berkeley of working with the Berkeley Labor Center, which brings together academics, students, labor and community organizers to tackle the challenges of linking education and research to projects of building worker and community power and effecting social change. It is a unique institutional space that I wish more universities would create. I would like to thank Katie Quan and Carol Zabin, who were the directors of the Labor Center while I was there, for always making me feel welcome; as well as the wonderful group of individuals working in and around the Center at the time, including: Jeremy Blasi, Sara Flocks, Nato Green, Lea Grundy, Pam Tau Lee, Scott Littlehale, Warren Mar, Peter Olney and Steve Pitts.

Since coming to the UCL Institute of Education in 2015, I have run an undergraduate placement module, *Educating and Organizing for Social Justice*, in partnership with Citizens UK, an affiliate of the Industrial Areas Foundation. This module has been a central place for working through many of the ideas discussed in this book. For working with me on this module and helping to shape my thinking and learning, I give special thanks to James Asfa and Hannah Gretton, both organizers with Citizens UK who have acted as module co-leads, their wonderful team of Citizens UK colleagues who have also worked on the module—including Claire Arkwright, Caitlin Burbridge, Jonathan Cox, Daphne Giachero, Tahmid Islam, Jessica Maddocks and Paulina Tamborrel Signoret—an inspirational group of school leaders from across London—including Natasha Beckles, Christine Bernard, Fiona Carrick-Davies, Mother Ellen Eames, Dawn Ferdinand, Rita Garcha, Susan Grace, Father Sebastian Harries, Bethan Tanner, Marta Tildesley and Matt Wright—and, of course, the many brilliant undergraduate students from UCL who have participated in this module.

While working at UCL, I have also experienced a very different kind of education that has likewise been central to shaping my thinking about the possibilities and challenges of social and educational transformation. Between 2018 and 2020, myself and my university colleagues have been out on extended national higher education sector strikes, not just once but three times. The UCL Institute of Education always has outstanding participation on its picket lines, which have been vital spaces for thinking about education for social change, through the pedagogy of the picket line itself, informal conversations with colleagues on the line, and daily Teach Outs that are organized on the line. For their friendship, solidarity and the deep learning we shared together during these strikes, I would like to thank: Theo Bryer, Amy Chamier, Jane Coles, Sheila Curtis, Annie Davey, Jay Derrick, Mark Newman, Rebecca O'Connell, Charlie Owen, John Potter, Mary Richardson, Holly Smith, Judith Suissa, Adam Unwin and John Yandell.

Finally, Mayssoun Sukarieh has played an integral role in shaping my thinking about the issue of radical education and social change; and some of the ideas in this book are things that we have written about together in previous publications. While I was working on this project, Mayssoun has been working on a parallel project exploring how traditions of radical education have been central to the ongoing liberation struggle in Palestine. As always, her work is an inspiration for myself and others to continue to seek to learn more about the many vital traditions of radical education that have been developed all around the world.

Contents

List of Figures

1
Introduction

As concern with global climate and environmental crisis has escalated, a call has appeared for new approaches to education that can facilitate the radical social, cultural and economic transformations across the planet that are deemed necessary to deal with this crisis effectively. In October 2018, the Intergovernmental Panel on Climate Change (IPCC) released its *Global Warming of 1.5 °C* report, that argued that limiting global warming to no more than 1.5 °C above pre-industrial levels—the threshold beyond which severe social, economic and environmental devastation caused by a warming climate was increasingly likely—would "require rapid, far-reaching and unprecedented changes in all aspects of society" (IPCC, 2018b). "Education, information and community approaches," the IPCC (2018a, p. 22) suggested, "can accelerate the wide-scale behaviour changes" needed for "adapting to and limiting global warming to 1.5°C." A few months earlier, a study published in the *Proceedings of the National Academy of Sciences* by an international group of climate scientists warned that "widespread, rapid, and fundamental transformations" in global society and economy will be essential to prevent extreme forms of climate change, and stated that we "need new collectively shared values, principles, and frameworks *as well as education* to support such

S. Tannock, *Educating for Radical Social Transformation in the Climate Crisis*, Palgrave Studies in Education and the Environment,
https://doi.org/10.1007/978-3-030-83000-7_1

changes" (Steffen et al., 2018, p. 6, emphasis added). UNESCO, which has adopted the slogan "changing minds, not the climate" for its campaign to confront the climate crisis, insists that "education is the most powerful element in preparing for the global challenges that climate change brings" and "a key enabler for a more sustainable future" (UNESCO, 2017, p. 2). In 2015, UNESCO Director-General Irina Bokova and UN Framework Convention on Climate Change Executive Secretary Christiana Figueres argued that "education can bring about a fundamental shift in how we think, act and discharge our responsibilities toward one another and the planet," and schools "can nurture a new generation of environmentally savvy citizens to support the transition to a prosperous and sustainable future" (Bokova & Figueres, 2015).

Calls for climate change education have not always had such a sense of urgency and radicalism. International declarations on climate change education, as Fumiyo Kagawa and David Selby (2015, pp. 33, 53) note, have often been characterized by "blandness" and reluctance to strongly question "business as usual." The authors point to the example of the *Lima Ministerial Declaration on Education and Awareness-Raising* from the 2014 United Nations Climate Change Conference in Peru, and its vague and tepid move to "encourage governments to develop education strategies that incorporate the issue of climate change in curricula," without specifying what such strategies could and should look like (p. 44). Indeed, for a long time, climate change education has been primarily approached as a form of science education, focused on increasing students' climate science literacy about the physical processes that are driving climate change (Busch et al., 2019; González-Gaudiano & Meira-Cartea, 2019; Henderson & Drewes, 2020). But as recognition has spread that addressing the climate crisis will require fundamental and rapid social change, a shift has occurred in how climate change education is talked about as well. There has been an acknowledgement that climate change education cannot be seen simply as a form of science education, but needs to be cross-disciplinary; and that it must be concerned not just with understanding the ways in which human societies have triggered a growing global climate and environmental crisis, but learning how to radically transform societies and economies to address this crisis as well. "'Climate change education' isn't simply 'climate education,'" Alan Reid

(2019, p. 768) writes. Climate change "is not just a scientific phenomenon" that "involves the natural sciences," argue Stevenson et al. (2017), p. 68), but a "complex socio-scientific issue that demands … educating for change [and] engaging the social sciences and humanities." Consequently, there been an explosion of research, theory and practice oriented literature that seeks to identify the most effective forms of educational intervention that can address the climate crisis and facilitate a radical and swift transformation toward worldwide sustainability.

This central question that is now being explored with a sense of urgency in the context of climate change education—whether and how education can be harnessed effectively for a broad-based, social justice inspired project of radical social transformation—is a vital one for all of us to consider, not just for addressing the climate crisis, but to tackle any number of other social, economic, political and environmental problems and crises as well. Currently, however, our ability to answer this question is often limited by two striking absences. First, despite the growing recognition in climate change discussions of the central role that needs to be played by education, educational researchers, theorists and educators have frequently been conspicuous by their absence from these discussions, as the field is dominated by research in psychology and communications. The IPCC *Global Warming of 1.5 °C* report, despite repeated mentions of the importance of learning and education in addressing the climate crisis, contains almost no references to work done by education researchers, drawing instead primarily on work done in the field of psychology. Anne Armstrong, Marianne Krasny and Jonathan Schuldt's (2018) book, *Communicating Climate Change: A Guide for Educators*, exemplifies this state of affairs, as it seeks to show climate change educators what they can learn about how best to do climate change education by drawing on "research from environmental psychology and climate change communication." As discussed in Chap. 3, the literature on climate change education has long been dominated by a behavioral science framework, that seeks to identify optimal conditions for fostering pro-environmental behavior change among individuals on a mass scale. There is no doubt that psychology and communications have much to offer for developing effective climate change education. But, equally, we are likely missing out by failing to look more closely at what the field of education theory, research and practice might have to offer as well.

Second, not only has the field of education research, theory and practice (beyond its subdisciplines of science education and environmental education) been slow to take up the challenge of addressing the climate crisis (Henderson et al., 2017); it has tended to neglect as well the broader question of how education can be used effectively to support projects of radical social change. The sociology of education, for example, has long been preoccupied with the question of how formal education works to prevent radical and progressive social change from occurring, by enabling the reproduction of structures of race, class and gender inequality in society (Gewirtz & Cribb, 2003). Works like *Education and Social Change* (Coffey, 2001), long used as an undergraduate education studies textbook in the UK, don't ask how education can lead to social change, but how education responds to and is impacted by social change that is occurring already. Much educational practice, theory and research is focused principally on supporting the ability of all children and learners to succeed and participate as equals in society as it now exists, rather than teaching them how to radically transform contemporary society. There is one vital area of education research, theory and practice where the question of how to use education to create a more just society has long been grappled with, drawing on different traditions of popular, progressive, feminist, anti-racist, anti-colonial, labor and democratic education (among others), but it has remained relatively marginalized in the field of education studies overall. This area is often referred to as the theory and practice of radical education (Fielding & Moss, 2010; Sukarieh & Tannock, 2016). Radical education, as Emily Charkin and Judith Suissa (2019, p. 394) write, may be defined as education that has an "underlying commitment to both critiquing the dominant norms, practices, values and institutions of existing society, and in positing an alternative." To be a radical educator, as Paulo Freire (2000, p. 37) suggests, is to approach education as a "process of liberation," one that is "nourished by a critical spirit" and committed to "ever greater engagement in the effort to transform concrete, objective reality."

The aim of this book is to start filling in these "striking absences" by bringing together work being done in climate change education with the broader and older traditions of radical education, to develop our understanding of how education can be, and often has been, used effectively to support collective projects of radical social transformation in the name of social justice. More specifically, this book seeks to address two core questions. What can we learn from current work being done on climate change education that can help us to better understand the possibilities and challenges of using education effectively for social justice inspired projects of radical social change more generally? How, in other words, can climate change education contribute to the larger field of research, theory and practice of radical education? Conversely, a second core question asked in this book is how the larger field of radical education can support and strengthen efforts to develop effective climate change education. Or, to pose the same question in a different way: what can the long, rich and diverse traditions of radical education offer current attempts to address the climate crisis, as one of the most essential issues of our time? In engaging with these questions, the book, in some chapters, seeks to pull together, build on and extend the links between these two bodies of educational work that a handful of scholars, researchers, educators and activists in the field of climate change education have already started to make. In other chapters, the links between climate change education and radical education that are explored are ones that, to a considerable degree, have been largely left missing in discussions of education and the climate crisis so far.

Climate Crisis, Education and Radical Social Change

Over the past few years, a small group of researchers and educators working in the field of climate change education have begun to explore some of the questions that are at the heart of this book. Asking what kinds of education are needed to help make the radical social, cultural and economic transformations that are essential for stopping and reversing climate change, this group calls for the embrace of "transformative" or "transgressive" learning and education (e.g., Boström et al., 2018;

Lotz-Sisitka et al., 2015; Macintyre et al., 2018). The ideas and arguments currently being proposed in this small, still marginal but growing literature are vitally important. Yet, in this book, rather than take this literature as a starting point, it serves instead as an endpoint, arrived at in the book's concluding pages. There are a couple of reasons for taking this approach. One is that there is a risk of tautology in calls for transformative education, as some of the group's authors themselves recognize. If what is needed is radical social transformation, then almost by definition, the kind of education that can support this will be a form of radical or transformative education. "This all sounds good and well," as Michael Peters asks one of the group's leading authors, "but what does social, transformative and transgressive education look like in practice?" (Peters & Wals, 2016, p. 186). After all, as Selby and Kagawa (2018, p. 306) warn:

> [The] ubiquitous, bandwagon usage of the notion of transformative learning carries its own downside. The language of transformation … 'can become so appealing it begins to be used for myriad purposes.' In other words, … it enters the realm of 'jelly words,' that is, terminology whose meaning becomes entirely malleable according to the agenda of the user. … [E]ven though sustainability educators frequently embrace the notion of transformative learning, there is little real consensus as to what sustainability-related education of transformative intent looks like.

A key task, therefore, is not just to invoke these broad labels of radical, transgressive or transformative education, but to start filling in the details of what these labels might mean in practice. How can we actually carry out an effective radical or transformative education that can help us address the climate crisis?

A second concern with some of the literature on transformative or transgressive climate change education is that it risks facing some of the same problems that have long faced critical pedagogy—which has been one of the more prominent streams of radical education in recent times, particularly in North American school-based education. Critical pedagogy has frequently been criticized for its inaccessibility to the very people it seeks to mobilize and empower, because it tends to favor highly theoretical, abstract and obscure language. As Barry Kanpol (1999,

p. 159) notes, "some students … rightly argue that critical pedagogy's obscure language makes radical educational ideas almost impossible to grasp" (see also Darder et al., 2009). Similarly, some texts in the transformative climate change education literature adopt this same genre of highly theoretical language, telling us, for example, that to develop effective climate change education that can support radical projects of social change, we need to embrace "(1) reflexive social learning and capabilities theory, (2) critical phenomenology, (3) socio-cultural and cultural historical activity theory, and (4) new social movement, postcolonial and decolonisation theory" (Lotz-Sisitka et al., 2015, p. 73). It is not that such arguments are necessarily wrong. On the contrary, many of these claims will be supported in this book, albeit in different ways and using different language. But to take this as a starting point requires a lot of time and intellectual energy to familiarize oneself with bodies of theory that are not always the easiest to understand, and work through how they may or may not be relevant and useful for effective climate change education practice.

The approach adopted in this book is to begin with relatively simple concepts that are regularly encountered in discussions of climate change education as key elements for doing this education effectively; and then, to turn to different traditions of radical education to explore how they can help to critique and develop these concepts in the places in which these concepts remain problematic, limited, contradictory or under-specified. The chapters are thus organized around concepts such as *hope*, *place*, *power*, *self interest*, *children and youth*, *curriculum*, and *nudging*. An advantage of this approach, it is hoped, is not only that most of these concepts are familiar and recognizable; but also that many educators, activists and researchers in the field of climate change are working with such concepts already, and thus might be able to see immediately what radical education traditions might offer for thinking through these concepts in alternative, more critical and transformative ways. The method for identifying the core concepts in this book was relatively simple: I selected key concepts that tend to appear frequently and/or prominently in popular and academic discussions about climate change education, and are concepts that radical education traditions have clear and important insights to contribute toward their effective development and operationalization. The list of concepts that runs through the chapter titles is not intended to be definitive or comprehensive; and we could easily think of other important concepts that might be additions or alternatives to the ones discussed here. However, all of the key concepts discussed in the chapters of this book point directly to theoretical arguments and practical, strategic engagements that stand at the heart of many, if not most, radical education traditions. The aim in focusing on these concepts is to initiate or push forward an engagement between climate change education and radical education traditions that will hopefully continue to develop in the future.

As this engagement between climate change education and radical education is developed in the following chapters, a number of patterns or themes emerge, about what it means to take a transformative or radical approach to climate change education. One is the movement beyond a concern with education as the development of knowledge about the climate crisis, to a recognition that education must also focus on the problem of learning how to take effective action that contributes to the radical social changes that are needed to address the crisis. A second is a shift away from a focus on how elite actors—government, business and civil society leaders, scientific experts, educational policy makers, and so forth—can reshape individual behavior toward the climate and environment, to ask how individuals, as citizens or community members, can come together to critically analyse and contest dominant social structures and cultural discourses, as well as the often problematic self interests and ideological agendas of elite actors in society. A third is the rejection of universalist and apolitical approaches to the climate crisis ("we're all in this together"), and insistence on recognizing the importance of deeply political, conflictual and situated differences in identity, power, status and location in shaping how different communities are impacted by and might be expected to take action (or not) to address the climate crisis immediately, as a matter of urgency. A fourth is the recognition that effective learning to address the climate crisis is not a narrowly academic, intellectual or cognitive matter, but one that is often deeply emotional and experiential, engaging all of our different senses. A fifth is the insistence that climate change education must encompass education in the broadest sense, to include not just schools, colleges and universities (formal education), but a broad range of nonformal and informal learning institutions and spaces as well. For those familiar with the different traditions of radical education, none of these themes will be a surprise. For all of them—the commitment to fostering education that can help to change and not just understand the world, the focus on structural analysis and collective rather than individual learning, the insistence on the inherent political nature of education, the embrace of holistic approaches to education, and the creation and use of informal and nonformal as well as formal education spaces—are, in many ways, part of what defines the core essence of radical education itself (see, for example, Fielding & Moss, 2010).

It is important to be clear about what this book does not aim to do. This is not a book that focuses on climate science or how to teach and learn about climate science. There are plenty of other good sources for these topics now. Rather, the focus here is on climate *change* education, or how to develop a form of education that can effectively support projects to radically transform contemporary society and economy, in the ways that are necessary to address the climate crisis. This book also does not attempt to make any argument about the specific kinds of social, cultural and economic changes that are needed to address the climate crisis. It may well be, as many now argue, that to address the climate crisis we need to move beyond capitalism as the prevailing social and economic system in the world, or beyond the obsession and structural imperative with fostering never-ending economic growth. It may be true, as well, that to address the climate crisis, we need to confront structural racism, patriarchy, colonialism and extractivism in society. However, while these and other such important arguments surface regularly throughout this book, the focus here is on education. The central question of concern is: what are the educational theories, strategies, practices, processes, institutions and locations that might be effective for enabling learners to engage directly with this kind of thinking about the necessity of and correct direction for radical social, cultural and economic change? Finally, this book is not intended to be a definitive or comprehensive overview of radical education, or the many different traditions—popular, progressive, democratic, feminist, anti-racist, anti-colonialist, and so on—that comprise the field of radical education as a whole. The strands of radical education that are engaged with are among some of the most influential, particularly in the western context. But there are many more vital radical education traditions beyond those discussed here, that we undoubtedly have much to learn from for tackling the climate crisis. The aim of this book is to contribute to and promote a necessary conversation about radical or transformative climate change education. But it certainly does not seek to have the final say on the subject.

Can Education Really Help to Change Society?

When asking how education can help support a broad based, social justice oriented project of radical social change, it can be useful to begin by paying attention to the claims of those skeptics who have long cast doubt on the possibility of education being able to lead to fundamental social change. Perhaps the most well known of these is Émile Durkheim's (1897/2005, p. 340) argument that looking to education to change society is "to ascribe to education a power it lacks." For Durkheim:

> [Education] is only the image and reflection of society. It imitates and reproduces the latter in abbreviated form: it does not create it. Education is healthy when people themselves are in a healthy state; but it becomes corrupt with them, being unable to modify itself. … Education … can be reformed only if society itself is reformed. (p. 340)

Durkheim's position presents a highly deterministic, derivative and uncontested account of schooling and education (Goldstein, 1976). But it also poses an important challenge to those of us hoping that education can help to foster radical social change, to be able to answer exactly how, why and where schools and other education institutions might be expected to acquire a transformative agency towards larger social and economic structures, within which they themselves are often deeply embedded. As Christina Kwauk and Olivia Casey (2021, pp. 63–64) note, as they argue for the promotion of a transformative climate change education, "the education system is characterized by a high degree of structural inertia and resistance to change," and any efforts to push for a "radical reorientation of the purpose and vision of education" risk "being avoided by teachers, rejected by ministries of education, and lobbied against by vested interests."

Many of the most important critiques of the limitations of education as a potentially transformative force have been made by scholars and educators who are themselves seeking to use education to change society. Zygmunt Bauman (2005, pp. 12, 14) notes that the "adverse odds" that stand against the transformational impacts of education are frequently "overwhelming," acknowledging that "the hopes of using education as a jack potent enough to unsettle and ultimately to dislodge the pressures of 'social facts' seem to be as immortal as they are vulnerable." In discussing the reasons for this vulnerability, Bauman makes the doubts expressed by Durkheim and others more concrete:

> The thrust of [radical, transformative] education … is to challenge the impact of daily experience, to fight back and in the end defy the pressures arising from the social setting in which the learners operate. But will the education and educators fit the bill? Will they themselves be able to resist the pressure? Will they manage to avoid being enlisted in the service of the self-same pressures they are meant to defy? (p. 12)

The "pressures" referred to by Bauman include prevailing cultural and ideological norms ("the ruling *doxa* and the daily evidence of commonsensical experience"), as well as mundane institutional constraints: school and university "boards of trustees," "government commissions," senior managers, and colleagues who prefer conformity, career security and advancement over the risks that come from "stirring the kids up" (p. 13). This is one of the most important questions confronting would-be radical, transformative educators hoping to address any number of social problems, including the climate crisis. How can an effective, transformative education be developed within schools and other education institutions, if the elite actors who oversee these institutions do not themselves already embrace such a radical ambition? Even if this challenge can be overcome, how can the radical lessons offered to students within schools and universities be retained and acted upon as students move into the realms of the workplace and wider civil society, where, again, dominant actors may be directly opposed to such radicalism? These are not hypothetical questions. In the United Kingdom, for example, the government recently moved to ban use of materials in schools that support calls to abolish or overthrow capitalism, critique white privilege or teach critical race theory, or more generally, promote "fundamental changes in political, economic or social conditions, institutions or habits of mind" (Sukarieh & Tannock, 2016, p. 30; Busby, 2020; Wood, 2020). Arguably, all of these are vital for any effective project to address the climate crisis.

George Counts, whose 1932 polemic, *Dare the School Build a New Social Order?*, is one of the best known and most influential calls to educators to embrace a transformative vision of education, begins his pamphlet with a critique and warning against the dangers of naïve and ungrounded faith in the power of education to solve social problems:

> Like all simple and unsophisticated peoples we Americans have a sublime faith in education. Faced with any difficult problem of life we set our minds at rest sooner or later by the appeal to the school. We are convinced that education is the one unfailing remedy for every ill to which man [*sic*] is subject, whether it be vice, crime, war, poverty, riches, injustice, racketeering, political corruption, race hatred, class conflict, or just plain original sin. We even speak glibly and often about the general reconstruction of society through the school. We cling to this faith in spite of the fact that the very period in which our troubles have multiplied so rapidly has witnessed an unprecedented expansion of organized education. This would seem to suggest that our schools, instead of directing the course of change, are themselves driven by the very forces that are transforming the rest of the social order. (Counts, 1932, p. 1)

Counts warns that education will only have a chance of contributing to radical social change if such naïve faith is questioned, so that "teachers must abandon much of their easy optimism" and "subject the concept of education to the most rigorous scrutiny" (p. 2). As discussed in Chap. 4, simply hoping that education will help us address the climate crisis, or any other social or environmental problem, will not make it so; and worse, could undermine the possibility of developing an effective transformative educational project and agenda.

Paulo Freire's critique of what he called "banking education" in the *Pedagogy of the Oppressed*, widely seen to be a criticism of traditional, mainstream education often found in schools and formal education institutions, was also directed at would be revolutionaries seeking to use education to radically transform society. Banking education is Freire's (2000, p. 73) term for a form of education conceptualized as "an act of depositing, in which the students are the depositories and the teacher is the depositor," and where the teacher "makes deposits which the students patiently receive, memorize and repeat." In banking education, according to Freire, "the teacher knows everything and the students know nothing," and "the teacher chooses the program content, and the students … adapt to it," and "comply" (p. 73). For Freire, banking education is an ineffectual, shallow form of education, and a disempowering, alienating and dehumanizing one. Freire warns that those who seek to radically change the world all too often embrace forms of banking education, and fatally undermine their own cause of liberation, empowerment and transformation:

> Unfortunately, those who espouse the cause of liberation are themselves surrounded and influenced by the climate which generates the banking concept, and often do not perceive its true significance or its dehumanizing power. Paradoxically, then, they utilize this same instrument of alienation in what they consider an effort to liberate. … Those truly committed to the cause of liberation can accept neither the mechanistic concept of consciousness as an empty vessel to be filled, nor the use of banking methods of domination (propaganda, slogans—deposits) in the name of liberation. (p. 79)

The general concern presented here is that educational projects for social change must pay close attention to matters not just of curriculum, but pedagogy and social relationships in educational institutions as well, if they are to have the possibility of success. More specifically, we also find here a warning against what is likely always going to be a strong temptation in developing climate change education, of presenting "the facts" to students—about climate science, the social and economic causes of climate change, and what needs to be changed in culture, society and the economy to tackle the climate crisis—as uncontested and uncontestable, universal and self-evident truths, and asking learners to sign up for a climate action agenda that has already been fully designed and planned out in advance.

Finally, John Dewey, who, like Counts and Freire, is one of the best known and most influential advocates of using education for social change, points to another set of concerns that must be kept in mind for projects for developing transformative approaches to climate change education (Pérez-Ibáñez, 2018). In a short essay on "Education and Social Change," Dewey argues that:

> It is unrealistic … to suppose that the schools can be a *main* agency in producing the intellectual and moral changes, the changes in attitudes and disposition of thought and purpose, which are necessary for the creation of a new social order. Any such view ignores the constant operation of powerful forces outside the school which shape mind and character. It ignores the fact that school education is but one educational agency out of many, and at the best is in some respects a minor educational force. (1937/1987, p. 414)

In asking the question of whether and how education can help to support a broad-based social justice oriented project of radical social change, it is essential that we adopt a wide understanding of and approach to education, as not just residing in schools, colleges and universities (formal education), but in countless other social settings as well, in organized, deliberate and collective forms and as an often unplanned and incidental consequence of everyday social experience (nonformal and informal education).

Many of the concerns expressed here about the possibilities for developing effective education for radical social change focus on challenges involved in trying to use schools, universities and other formal education institutions for such projects—and this implies that effective climate change education might best be explored outside formal schooling. There is no doubt that broadening our understanding of what is meant by education is essential, and making use of all kinds of spaces for learning is vital. Furthermore, history shows us that many of the most successful and influential educational spaces for radical social change are ones that have been deliberately constructed outside the school system, as discussed in Chap. 8. But to turn our back on schools and other formal education institutions as key sites for developing a transformative climate change education would be a mistake, even if these can often be extraordinarily difficult places in which to make much headway.

Rebecca Tarlau's recent work on the Landless Workers Movement's longstanding effort to engage directly with state education systems in Brazil offers important insights into just why this engagement is so important. Even though the Landless Workers Movement is a well established social movement in its own right that regularly creates and runs its own education institutions, projects and spaces, it seeks—despite many obstacles and frequent setbacks—to work with and reshape state run school systems in Brazil as well. One reason for this is the importance of outreach. As Tarlau (2019, p. 40) notes:

> [M]ovements' participation in public schooling can help recruit new activists, and in particular, youth and women to the movement. Public schools and universities are institutions where young people spend many hours each day and are therefore important spaces for investing in local leadership development—increasing students' interest in and capacity for social change. … [D]iscussions [about social change] can also happen in other social-movement-led training programs, such as popular education. However, the people who participate in these nonformal educational spaces are often already active supporters or at least sympathetic to the movements. … A social movement's participation in formal schooling can help convince youth, who may have never participated in a contentious protest, to become involved in these collective struggles.

Other key factors pointed to by Tarlau are the fact that schools and formal education institutions usually have access to far more "financial and institutional resources" than are commonly available to non-formal educational spaces set up by social movements; and that "if movements can develop a degree of influence in the public schools," then these schools can be "important locations where social movements can begin to prefigure, in the current world, the social practices that they hope to build in the future," across the rest of society and the economy as well (pp. 40, 42).

In the end, the question about where we should be attempting to develop radical, transformative forms of climate change education needs to be framed not as an "either-or" choice between formal versus nonformal sites of education, but a "both-and" matter that recognizes the potential complementarity of these different spaces of learning. Indeed, the question of "how different educational practices could and should be distributed across different educational spaces," including not just schools and universities, but other institutions such as "political parties, trade unions, social movements, community groups, churches and other faith-based organizations" should be seen a central concern "of both principle and strategy" for all educational projects for radical social change to consider (Tannock et al., 2011, p. 942).

References

Armstrong, A., Krasny, M., & Schuldt, J. (2018). *Communicating climate change: A guide for educators*. Cornell University Press.

Bauman, Z. (2005). *Liquid life*. Polity Press.

Bokova, I., & Figueres, C. (2015). Why education is the key to sustainable development. *World Economic Forum*, 19 May. https://www.weforum.org/agenda/2015/05/why-education-is-the-key-to-sustainable-development/

Boström, M., et al. (2018). Conditions for transformative learning for sustainable development: A theoretical review and approach. *Sustainability, 10*(4479), 1–21.

Busby, M. (2020). Schools in England told not to use material from anti-capitalist groups. *Guardian*, 27 September. https://www.theguardian.com/education/2020/sep/27/uk-schools-told-not-to-use-anti-capitalist-material-in-teaching

Busch, K., Henderson, J., & Stevenson, K. (2019). Broadening epistemologies and methodologies in climate change education research. *Environmental Education Research, 25*(6), 955–971.

Charkin, E., & Suissa, J. (2019). Radical education. In R. Kinna & U. Gordon (Eds.), *Routledge handbook of radical politics* (pp. 391–404). Routledge.

Coffey, A. (2001). *Education and social change*. Open University Press.

Counts, G. (1932). *Dare the school build a new social order?* Southern Illinois University Press.

Darder, A., Baltodano, M., & Torres, R. (2009). Critical pedagogy, an introduction. In A. Darder, M. Baltodano, & R. Torres (Eds.), *Critical pedagogy reader* (pp. 1–20). Routledge.

Dewey, J. (1937/1987). Education and social change. In J. Boydston (Ed.), *Later works of John Dewey* (Vol. 11, pp. 408–418). Southern Illinois University Press.

Durkheim, E. (1897/2005). *Suicide*. Routledge.

Fielding, M., & Moss, P. (2010). *Radical education and the common school*. Routledge.

Freire, P. (2000). *Pedagogy of the oppressed*. Continuum.

Gewirtz, S., & Cribb, A. (2003). Recent readings of social reproduction: Four fundamental problematics. *International Studies in Sociology of Education, 13*(3), 243–260.

Goldstein, M. (1976). Durkheim's sociology of education: Interpretations of social change through education. *Educational Theory, 26*(3), 289–297.

González-Gaudiano, E., & Meira-Cartea, P. (2019). Environmental education under seige: Climate radicality. *Journal of Environmental Education, 50*(4), 386–402.

Henderson, J., & Drewes, A. (2020). Teaching climate change in the United States. In J. Henderson & A. Drewes (Eds.), *Teaching climate change in the United States*. Routledge.

Henderson, J., et al. (2017). Expanding the foundation: Climate change and opportunities for educational research. *Educational Studies, 53*(4), 412–425.

IPCC—Intergovernmental Panel on Climate Change. (2018a). *Global warming of 1.5°C*. IPCC.

IPCC—Intergovernmental Panel on Climate Change. (2018b). Summary for policymakers of IPCC special report on global warming of 1.5°C. 8 October. https://www.ipcc.ch/2018/10/08/summary-for-policymakers-of-ipcc-special-report-on-global-warming-of-1-5c-approved-by-governments/

Kagawa, F., & Selby, D. (2015). The bland leading the bland: Landscapes and milestones on the journey towards a post-2015 climate change and development agenda and how development agenda can reframe the agenda. *Policy & Practice, 21*, 28–57.

Kanpol, B. (1999). *Critical pedagogy: An introduction*. Bergin & Garvey.

Kwauk, C., & Casey, O. (2021). *A new green learning agenda: Approaches to quality education for climate action*. Brookings.

Lotz-Sisitka, H., et al. (2015). Transformative, transgressive social learning: Rethinking higher education pedagogy in times of systemic global dysfunction. *Current Opinion in Environmental Sustainability, 16*, 73–80.

Macintyre, T., et al. (2018). Towards transformative social learning on the path to 1.5 degrees. *Current Opinion in Environmental Sustainability, 31*, 80–87.

Peters, M., & Wals, A. (2016). Transgressive learning in times of global systemic dysfunction: Interview with Arjen Wals. *Open Review of Educational Research, 3*(1), 179–189.

Pérez-Ibáñez, I. (2018). Dewey's thought on education and social change. *Journal of Thought*, 52(3), 19-31.

Reid, A. (2019). Climate change education and research: Possibilities and potentials versus problems and perils? *Environmental Education Research, 25*(6), 767–790.

Selby, D., & Kagawa, F. (2018). Teetering on the brink: Subversive and restorative learning in times of climate turmoil and disaster. *Journal of Transformative Education, 16*(4), 302–322.

Steffen, W., et al. (2018). Trajectories of the earth system in the Anthropocene. *Proceedings of the National Academy of Sciences, 11533*, 8252–8259.

Stevenson, R., Nicholls, J., & Whitehouse, H. (2017). What is climate change education? *Curriculum Perspectives, 37*, 67–71.

Sukarieh, M., & Tannock, S. (2016). The deradicalisation of education: Terror, youth and the assault on learning. *Race & Class, 57*(4), 22–38.

Tannock, S., James, D., & Torres, C. (2011). Review of *Radical education and the common school. British Journal of Sociology of Education, 32*(6), 939–952.

Tarlau, R. (2019). *Occupying schools, occupying land: How the Landless Workers Movement transformed Brazilian education*. Oxford University Press.

UNESCO. (2017). *Changing minds, not the climate: The role of education*. UNESCO.

Wood, V. (2020). Teachers presenting white privilege as fact are breaking the law, minister warns. *Independent*, 22 October. https://www.independent.co.uk/news/uk/politics/kemi-badenoch-black-history-month-white-privilege-black-lives-matter-b1189547.html

2

Curriculum Struggles: Knowledge, Truth … Action?

"Our learning needs to reflect the severity of the climate crisis," states a group called Teach the Future (2020d): "We demand reform to the education system." Created out of the school student climate strike movement that swept across the United Kingdom in 2019, Teach the Future is a youth-run campaign to further the demands of student strikers for schools, colleges and universities in the UK to do more to address the climate crisis—or, as the group puts it, "to urgently repurpose the entire education system around the climate emergency and ecological crisis." Teach the Future is backed by an array of groups, including 350.org, Friends of the Earth, Greenpeace, Extinction Rebellion, the National Education Union, and Royal Society for Arts. In February 2020, Teach the Future held an event at the Houses of Parliament in London, where a group of forty seven students aged 13 to 26 presented their arguments to a cross party group of over fifty MPs (Ibrahim, 2020; Tickle, 2020). At the center of their demands was a request for widespread curriculum change. Schools don't do enough to teach about climate and environmental crisis, the group argues, pointing to surveys in the UK that found that just 4% of primary and secondary students feel "they know a lot about climate change," 42% of secondary students "say they have learnt

S. Tannock, *Educating for Radical Social Transformation in the Climate Crisis*, Palgrave Studies in Education and the Environment,
https://doi.org/10.1007/978-3-030-83000-7_2

a little, hardly anything or nothing about the environment at school," and "68% want to learn more about the environment and climate change" (Teach the Future, 2020a). Much of what is taught at school and university, the group claims, "is irrelevant to the uncertain future we are inheriting" and "out of touch with the future we face" (Teach the Future, 2020c). Since "current and future generations of students are growing up in a world shaped by the climate crisis," Teach the Future calls on the government to urgently commission a comprehensive review of how the "education system is preparing students for the climate emergency and ecological crisis," and pass a Climate Emergency Education Act that, among other things, would require "all education providers to teach the truth about the climate and ecological emergency" (Teach the Future, 2020b, 2020e).

Curriculum struggles are often at the heart of mobilizations around climate change education, as they are for many other social justice causes. Whether our concern is racism, colonialism, sexism, poverty, inequality, ablism, homophobia, public health, drug use, and so on, a natural starting point is to seek to get these issues covered in the school and university curriculum (Labaree, 2008). "The formal curriculum has been considered critically significant by social movements of all kinds," Julie Gorlewski and Isabel Nuñez (2020, p. 4) note, "and passionate debates over the in-school curriculum have been at the heart of some of the most contentious feuds in education." There is little mystery about why there is so much focus on curriculum. For the official curriculum "lies at the heart of schooling communicating the most important messages to youth about what we value and why we educate," about what "orientations are deemed valuable and made readily available to students," and about "what is perceived to be 'legitimate knowledge'—the knowledge that 'we all must have'" (Apple & Franklin, 2019, p. 65; Au, 2012, p. 51). If there is a crisis that threatens life as we know it, then surely it is a matter of urgency to ensure this is addressed centrally and substantively by the school curriculum.

At the same time, however, there have long been questions about how impactful changes to the school and university curriculum actually are. As Larry Cuban (1993, p. 192) writes, in a widely cited article on "the lure of curricular reform and its pitiful history":

> The most common strategy that reformers have used … to get students to know and do the right things is to change the curriculum. That popular strategy has largely failed. If this history of failure were to be known more widely, it might embarrass the present generation of reformers who scale steep hills to plant the flag of curriculum reform. It is humbling to realize how little each generation learns from the experience of its equally earnest forebears about just how crude a tool curriculum change is for transforming student knowledge and behavior.

While the state and shape of the formal curriculum in schools and universities is politically and symbolically important, it remains unclear how much this curriculum changes in student learning and practice, let alone in the wider world beyond.

This chapter reviews some of the recent battles and struggles over climate change curriculum, before pointing to some of the problems and limitations of curriculum focused efforts to tackle the climate crisis. These include the need to address school structure and practice, or what is called the core grammar or hidden curriculum of schooling; the need to overhaul our entire way of thinking about climate and environmental crisis, by shifting away from projects that seek to add units to pre-existing school curricula, and toward a more fundamental paradigm shift in how we build our understanding of the social and physical worlds in which we live; and the need to rethink the model of social change that places curricular knowledge at the center of radical, transformative social action.

School Curriculum as Climate Battleground

Truth in Textbooks is a conservative citizens group based in Texas that started out in 2013 fighting what it saw as anti-Christian and anti-American bias in school textbooks (Foran, 2014; TNT, 2020). Offering a three month training course for volunteers to learn to do line by line reviews of school textbooks in the hunt for offending passages, produce a rating system of school textbooks according to how truthful they are (or not), and lobby for the revision or removal of problematic textbooks, Truth in Textbooks quickly expanded its operations to twenty other states across America, and began to focus on the issue of how climate change was being taught in US schools. The enemy was what the group identified as "a one-sided global-warming climate-change agenda," that seeks to teach that human-induced climate change is an accepted reality as opposed to contested opinion, and threatens to instill "fear in children at a very young age that either we're going to run out of something or over-pollute the Earth," thus giving them "the wrong impression of America" (Foran, 2014; Holland, 2020).

In 2017, Truth in Textbooks linked up with a similar group called the Florida Citizens Alliance to provide its textbook review training course to volunteers in Florida. The Florida Citizens Alliance had launched a successful campaign to get Florida to enact the country's first state bill giving every resident the right to make official complaints about textbooks and other materials used in local schools they find to be objectionable, and requiring school districts to hold public hearings to investigate and respond to such complaints formally (Weyrauch, 2017; Worth, 2017). For the Alliance, this was about taking back public control of schools from perceived domination by liberal elites. "We're really concerned," one Alliance member stated, "that our kids are not being educated, [but] … indoctrinated in the philosophy of the academic aristocracy" (Cooper, 2018). Like Truth in Textbooks, the Florida Citizens Alliance opposed the teaching of climate change in state schools, claiming that "man-made global warming is a hoax" and "pure and unadulterated false propaganda." Their partnership with the Texas based outfit was an attempt to improve their effectiveness at using the new Florida law to promote this and other political and educational agendas (Otto, 2017). The irony is that this is taking place in Florida, a state argued to be among the most imminently threatened by the rising sea levels and increased prevalence of severe hurricane weather brought about by a rapidly changing climate (Dawson, 2017).

Truth in Textbooks, the Florida Citizens Alliance and the Florida textbook law are all part of a nationwide battle over climate change curriculum in schools that has continued in the United States over the past two decades and more. In the five years between 2015 and 2020, bills have been introduced in at least eighteen states that seek to restrict or roll back climate change education in public schools (Bagley, 2015; Czajka, 2019;

Firozi, 2019; Melia, 2019; Worth, 2017). In Idaho, a political tug of war has gone back and forth throughout this period over whether or not to include climate education in state school science standards (Albeck-Ripka, 2018; Worth, 2019). Every time climate education supporters have managed to get climate education into the standards, Republican lawmakers have fought back. "By teaching proper science," one Republican Representative argued in January 2020, as he opposed climate change education before the state's House Education Committee, "we can even improve the mental health of some of the Idaho school children ... by removing unfounded fears and preventing the Greta Thunberg syndrome" (Dawson, 2020). In 2017, the conservative Heartland Institute mailed a book called *Why Scientists Disagree About Global Warming* to 350,000 school and college science teachers across the United States (Worth, 2018b). Many of the groups engaged in fighting against climate change education are funded and supported by the fossil fuel industry, or motivated by the extensive networks of climate change denial and obfuscation that have been set up by this industry over the last few decades (Mann & Toles, 2016; Mann, 2021).

Political mobilization around climate change education is not all in one direction. In Texas, the Texas Freedom Network challenges the arguments of Truth in Textbooks, while in Florida, a group called Florida Citizens for Science tries to act as a brake on the anti-climate education agenda of the Florida Citizens Alliance (Cooper, 2018; Foran, 2014). The National Center for Science Education monitors anti-climate change education legislation and action across the United States, and supports teachers, schools and local community members in developing, promoting and protecting effective climate change curriculum. In October 2020, the Center released its *Making the Grade?* Report, that ranked school science standards in all fifty states on a scale of A to F for how well or poorly they promoted climate change education (NCSE & TFN, 2020). In Ithaca, two groups at Cornell University sought to counter-balance the efforts of the Heartland Institute by mailing out a book, *The Teacher-Friendly Guide to Climate Change*, to schools across the US, and making another book, *Communicating Climate Change: A Guide for Educators*, free to download (Worth, 2018a). In 2018, the National Science Teachers Association released a statement on "the teaching of climate science," raising concern that "teachers are facing pressure to not only eliminate or de-emphasize climate change science, but also to introduce non-scientific ideas in science classrooms," and calling on policy makers to "preserve the quality of science education by rejecting censorship, pseudoscience, logical fallacies, faulty scholarship, narrow political agendas, or unconstitutional mandates" (NSTA, 2018). In 2013, a coalition of states, along with the National Research Council, National Science Teachers Association, American Association for the Advancement of Science and Achieve, released the Next Generation Science Standards, that set high quality, research-driven standards for science education that could be adopted nationwide, and promoted the adoption of rigorous climate change education as a core part of the school science curriculum (Holland, 2020; NCSE & TFN, 2020).

Other countries haven't seen quite the same level of overt political fighting over the climate curriculum, but it does exist. In December 2020, in the United Kingdom, the Chief Inspector of Schools in England accused climate change activists who were calling for "new qualifications or more explicit alterations to the curriculum" of engaging in an effort "to commandeer schools and the curriculum" and turn science education into "a morality tale or something quasi-religious." The Chief Inspector insisted we should not "revise the curriculum in the context of a single issue or purpose" (Turner, 2020). Earlier the same year, a Conservative MP introduced an Anxiety in Schools (Environmental Concerns) Bill to UK Parliament, warning that "extremist ideologies" embraced by climate activist groups like Extinction Rebellion were having "a very serious impact upon the ignorant, upon the naïve," and upon "tender minds" (TES, 2020). In 2013, a move by the UK government to remove climate change from the national science curriculum—allegedly to "placate the 'right wing of the Conservative party'" through omitting discussion of "'the human causes' of climate change"—led to a mass outcry and mobilization, with 65,000 people signing petitions of protest (BBC, 2013; Coughlan, 2017).

In Australia, conservative governments "have shown no interest in making climate change a key education issue, and have been vocal about keeping 'activism' and 'politics' out of schools" (Climate Scorecard, 2020, p. 4). In 2019, the Education Council of state governments adopted "a new national declaration on education goals for all Australians" that removed all references to climate change that had existed in previous declarations (Gough, 2020a). A conservative group called Advance Australia launched a campaign in 2020 to send educational resources to primary schools across Australia that aimed to challenge the "climate alarmist narrative" and "politically correct indoctrination and groupthink" it believes is prevalent in the education sector (Donnelly, 2020; Koziol, 2020). These resources include *The Smart Scientist's Kit*, that explains to children "how Australia is blessed with an excess of natural resources and that it doesn't make sense to try and stop the climate from changing" (Koziol, 2020). In Canada, much of the political fighting around climate change education has occurred in the oil sands province of Alberta. In 2017, the provincial premier attacked a proposed change to the Alberta school curriculum to focus on the issue of climate change (Lawrynuik, 2019). In 2020, the education minister criticized Alberta schools for teaching "the 'extremist view' that we're running out of time to get the climate crisis under control," and insisted we need to "get politics out of the classroom." An education panel convened by the provincial government proposed that the school curriculum should highlight "the importance of Alberta's resource-rich economic base [in other words, the oil sands] in relation to the impact on the economy, families, services, and government" (Energy Mix, 2020; Zabjek, 2020).

Elsewhere, the struggle to develop and disseminate climate change curriculum throughout regional and national educational systems is less confrontational and overtly partisan political. States and educational authorities often have other educational concerns that take priority over the promotion of climate change education. As Christina Kwauk (2020, p. 9) observes, in her review of global climate education:

> Low- and middle-income countries ... face a double burden in the context of climate change and sustainable development. Many of these countries are highly vulnerable to weather-related disasters and the negative impacts of a changing climate. At the same time, many are struggling to deliver basic education services, to address the learning crisis, to include girls, refugees, and other marginalized or minority communities, and a host of other education-related development challenges. Addressing climate change ...

> becomes another checkbox on a long list of priorities for governments, civil society, implementers, and donors.

Many states and educational authorities in the contemporary era are committed to social, economic and educational agendas that conflict with or undermine effective climate change education—such as the commitment to unfettered economic growth, neoliberal, free market ideologies, and the vocationalized prioritization of job training and labor force skills development. Some states and educational authorities may lack the necessary resources or centralized power structures that would aid the effective promotion and support of climate change education at scale (Climate Scorecard, 2020; Henderson & Drewes, 2020). One of the most basic struggles in getting schools to provide effective climate change curriculum, Alan Reid (2019, p. 770) suggests, is that "local and national [education] sectors and provisions are so often ill-thought, underfunded, overstretched, and undervalued already."

In all of these struggles, the central issue is not just about whether to teach about climate change as a human induced phenomenon, but how to teach about it, as well as how much to teach about it. Over time, promotion of direct climate change denialism in schools by the fossil fuel industry and its supporters has become less prominent, as these groups instead embrace what Michael Mann (2021, p. 3) calls "a softer form of denialism." Hard-core climate deniers, Mann warns:

> are being replaced by other breeds of *deceivers* and *dissemblers*, namely, *downplayers, deflectors, dividers, delayers,* and *doomers*—willing participants in a multipronged strategy seeking to deflect blame, divide the public, delay action by promoting 'alternative' solutions that don't actually solve the problem, or insist we simply accept our fate—it's too late to do anything about it anyway, so we might as well keep the oil flowing. (p. 46; italics in original)

Rather than opposing climate change education outright, this new set of "soft denialist" strategies seeks instead to shape the ways in which climate change is taught. This includes calls to provide a "balanced" curriculum that teaches "both sides" of the argument about human-induced climate change being real or not, based on a claim that the existence of such change remains scientifically contested (despite there being no evidence that this is the case) (Energy Mix, 2020; Koziol, 2020).

Fundamentally, these strategies promote what Emily Eaton and Nick Day describe as "petro-pedagogy," a form of education that continues to serve fossil fuel industry interests. One prevalent form of petro-pedagogy, in Eaton and Day's (2020, p. 458) analysis, adopts a model of "neoliberal environmentalism" that restricts "the imagination of possible climate solutions to individual acts of conservation," thus failing "to challenge the structural growth of fossil fuel production" and working to "insulate fossil fuel industries from criticism and dissuade young people from questioning or understanding the role of corporate power in the climate crisis." Perhaps the most clear-cut example of this is the extensive promotion by BP and public relations agency Ogilvy and Mather in the early 2000s of the personal carbon footprint calculator, that allows individuals to calculate the total amount of greenhouse gas emissions they produce each year, just by going about their routines in daily life (Kaufman, 2020). Now widely found in climate change education classrooms and programs around the world, the personal carbon footprint calculator has the intention and effect of entrenching "the perception that the responsibility for reducing [carbon] emissions [lies] with individuals," and not with fossil fuel corporations such as BP, a company which has long been "near the top of the list of the highest-emitting companies in the world" (Yoder, 2020). While "individual efforts to reduce one's carbon footprint are laudable," argues Mann (2021, p. 81), "without systemic change, we will not achieve the massive decarbonization of our economy that is necessary to avert catastrophic climate change."

Getting Stuck in the Core Grammar of Schooling

As important as these struggles over promoting climate change curriculum in schools are, their focus has tended to remain relatively narrow. These struggles have mostly been about trying to get climate change education included as part of the official science curriculum in regional and national school curriculum standards, and to get good climate science education resource materials into schools. This leaves open the question of how standards are translated and enacted, and how resources are used, in actual practice at the classroom level (Branch et al., 2016; Colston & Ivey, 2015). One key issue here concerns the recruitment, training and support of teachers who are able to deliver an effective climate change curriculum to their students. Some teachers may lack knowledge of how to teach about climate change, some may come from backgrounds that make them less inclined to see climate change as a pressing concern, and teacher training and professional development on this topic remains limited in many countries (Drewes et al., 2018; Kwauk, 2020; NCSE, 2016). It is not just curricular and pedagogical training and support that teachers may need. Researchers have found that teachers sometimes express concern about student, parent and community indifference or resistance to learning about climate change, and limit or adjust their teaching about the issue accordingly (Colston & Vadjunec, 2015; Monroe et al., 2019). Teachers often seek to balance the need to enact "policies set by the state and federal governments," while recognising that "they also serve the local public and are held accountable if they diverge too far from the expectations of their local communities" (Branch et al., 2016, p. 91).

This is not a challenge particular to climate change education, but one that has long been recognised in radical forms of social justice education. It is about addressing the question of Zygmunt Bauman (2005, p. 12), cited in the Introduction, of whether educators will be able "to challenge the impact of daily experience, to fight back and in the end defy the pressures arising from the social setting in which … learners operate." "Can you expect teachers to revolutionize the social order for the good of community?," asked Carter Woodson (1933, p. 68) in *The Mis-Education of the Negro*, as he sought to use education to challenge race inequality and injustice in the United States in the early twentieth century. "Indeed we must expect this very thing," Woodson argues, since "the educational system of a country is worthless unless it accomplishes this task" (p. 68). But if teachers are to be able to take on such roles and "become change agents," they need to be provided with extensive, continuing "in-class support, financial and pedagogical resources, creative space, and professional development opportunities" (Kwauk, 2020, pp. 15–17). Woodson thus set about building an extensive network of institutions, programs, resources and other supports—including "Negro History Week, the *Negro History Bulletin*, his black history textbooks, and ASNLH [the Association for the Study of Negro Life and History]"—that could assist and empower African-American teachers throughout the United States in "the task of implementing a corrective and transformative black education" (King et al., 2010, p. 214; see also Apple, 2013; Givens, 2021). Woodson's legacy offers one model for mass teacher engagement and radicalization that climate change educators and activists could learn from now.

Beyond the issue of teacher recruitment, training and support, a central observation of critical curriculum studies is that the official curriculum constitutes only one part of what is learned by students in schools, and is often not even the most consequential. Larry Cuban (1993) distinguishes official curriculum from the "taught curriculum"—noting that different teachers will often teach the "same" curriculum in different ways—as well as the "learned curriculum"—which includes the tacit, informal learning of students in schools—and the "tested curriculum"—that which schools actually assess students on. Michael Apple (2019, p. 13) highlights the importance of the "hidden curriculum," defined as:

> the tacit teaching to students of norms, values, and dispositions that goes on simply by their living in and coping with the institutional expectations and routines of schools day in and day out for a number of years.

This concept is discussed in greater detail in Chap. 3. Wayne Au (2012, p. 38) argues for a comprehensive understanding of curriculum as "a problem of complex environmental design," that recognizes that "the structure of the educational environment around the learner has implications for the acquisition of knowledge by that learner." On this account, student learning is shaped by physical materials in schools (including furniture, equipment, architecture), language and symbol systems, social relations between students, teachers and other staff, and how school time is organized (Au, 2012, pp. 35–37). David Tyack and William Tobin (1994, p. 454) likewise argue there is a "basic 'grammar' of schooling" that shapes student learning, often impeding the effectiveness of curriculum reform efforts, they identify as "the regular structures and rules that organize the work of instruction," including "standardized organizational practices in dividing time and space, classifying students and allocating them to classrooms, and splintering knowledge into 'subjects.'" The risk is that even if climate change becomes a topic in the official curriculum, it may be that student learning is not changed much, as the direction and nature of this learning is strongly shaped by structural characteristics of the overall school learning environment.

The dilemma for those seeking to use school climate change curriculum to address the climate crisis is one that has long been recognized by progressive, radical, democratic and feminist educators. In the field of environmental education, it was raised in a widely cited article by Robert Stevenson (1987) thirty years ago: there are often "contradictions in purpose and practice" between the radical, transformative aims of social justice curriculum and the conservative learning environment reproduced by the basic grammar of schooling. For Stevenson, the "first major contradiction" is that schooling is oriented to "conserving the existing social order," through reproducing prevailing "norms and values" in younger generations, and has not been organized with "the revolutionary purpose of transforming the values that underline our decision making" (p. 145). School learning "tends to be atomistic and individual," rather than "holistic and co-operative," as it is shaped by "the school's role in credentialling students and determining their future opportunities by means of competitive grading and ranking" (pp. 147–148). For students, school learning often has an exchange value that can help them enhance their "individual status and economic well-being," more than a "social value," that can enable them to work collectively to create "a sustainable and emancipated quality of life" for all (p. 147; see also Hursh et al., 2015). Rather than environmental and climate change education transforming the traditional structures and purposes of schooling, the core grammar of schooling can end up disciplining and domesticating these kinds of transformative educational initiatives (Gruenewald & Manteaw, 2007). Even if activists succeed in getting climate change into the curriculum, it can become just another curriculum unit for students to be tested on, with good grades in this unit, like all others, sought after by students not so much so they can address the climate crisis directly, but to increase their chances of future entry into top universities and well paid jobs.

If we are going to look to schools to help address the climate crisis, there is a need to consider more than just the formal curriculum. To be fair, groups like Teach the Future recognise this. Teach the Future (2020a) asks that all new and current teachers in the UK are taught about the climate crisis; and that a "Climate Education Information Institute" be created that can provide up to date "teaching resources as well as continuous professional development (CPD) materials." Recognizing that "students learn from their environment," Teach the Future (2020a) asks that "education buildings should [become] environmentally friendly, with students engaged in the changes that are made to their buildings." Agreeing with Stevenson's (1987) concerns that school learning tends to be oriented to exams, grades and credentials as primary educational outcomes, Teach the Future (2020a) asks for the creation of a "Youth Climate Endowment Fund" that can "fund youth-led climate and environmental social action in every educational institution to engage their peers, teachers and parents in practical, local action to abate the climate emergency and ecological crisis." Finally, Teach the Future (2020d) calls for an urgent review that can work to "repurpose the entire education system around the climate emergency and ecological crisis." These are obviously much bigger demands than adding climate change to the curriculum, as they are about comprehensive restructuring of schooling as a whole. The dilemma is that the bigger these educational asks and the more structural these requested changes in schooling become, the harder it is to be successful in getting such demands approved and enacted throughout regional and national education systems.

Recognizing the limitations of relying on the formal school curriculum to facilitate effective climate change education and action, UNESCO calls for the embrace of "whole-school approaches to climate action." As UNESCO (2016, p. 3) explains the concept:

> In a whole-school approach, students' classroom learning about climate change is reinforced by the formal and informal messages promoted by the school's values and actions. In other words, students ... and other members

> of the school community live what they learn, and learn what they live. The whole-school approach to climate change means that an educational institution includes action for reducing climate change in every aspect of school life. This includes school governance, teaching content and methodology, campus and facilities management as well as cooperation with partners and the broader communities.

The whole school approach seeks to tackle the problems identified by critical curriculum scholars as the principal sources of curriculum reform failure, by transforming not just the formal curriculum but the entire, complex educational environment that surrounds each individual learner within any school (Wals & Benavot, 2017). In particular, it attends to the central importance of informal learning by students as well as formal learning—an issue discussed in more detail in Chap. 3. The approach has led to a remarkable proliferation of Eco-Schools, Green Schools and Sustainable Schools in countries around the world—a so-called "green school movement"—that attempt to reorient everything these schools do around a core commitment to addressing environmental and climate crisis (Chopin et al., 2018; Gough et al., 2020; Hargis et al., 2018).

While promising, the whole school approach has its own challenges and limitations. One is a problem that has long been faced by progressive school reformers. Due to the lack of full support from regional and national education ministries, it is not the entire education system that is being transformed by the green school movement, but a minority of schools. Annette Gough (2020b, pp. 423, 427) estimates that "generally a third or less" of schools in any given country are currently participating in green school programs, "with a domination of early childhood and primary schools." Unless the best Green Schools and Eco-Schools can become prefigurative spaces for developing pilot models that lead to the transformation of the whole school system, the risk is that this kind of climate and environmental education can end up becoming what so often happens to progressive forms of education: a specialist form of schooling accessible only to a privileged few children and young people, who often come disproportionately from middle class and elite backgrounds (e.g., Beard, 2020; Chertoff, 2013).

In summarizing a review of Green Schools in twenty countries around the world, Gough (2020b) further points out that, as with any such labels, the moniker of being a "Green School" can obscure considerable differences in theory, structure and practice. In some Green Schools, Gough notes, teachers are unwilling "to embrace new pedagogical practices, in particular a socially-critical pedagogy, and so achieve socially-transformative education"; and not all Green Schools, despite their name, truly "adopt a whole school approach" (p. 427). There are essential differences that need to be acknowledged between Green Schools that accept corporate funding and sponsorship and those that do not (Huckle, 2013); or between a Green School such as Green School Bali, that caters to international students who fly into this private residential school from around the world, and whose parents can afford the annual school fees of \$12,000–\$24,000 US dollars, and a Green School that is a public day school with no fees, that serves children from its surrounding community who come from a range of class backgrounds (Green School Bali, 2020; Vanderbilt, 2018).

While the whole school approach may help to address the problem of conflicts and inconsistencies between the formal curriculum and the educational environment provided by any given school, its wider significance depends on the overall vision that is promoted within each whole school approach. One concerning finding is that at least some Green Schools focus on promoting individual lifestyle change and personal responsibility as primary solutions to climate and environmental crisis, and neglect what Kwauk and Casey (2021, p. 9) refer to as "transformative learning": learning to analyse the role of social and economic systems in creating the climate crisis, and shaping the unjust and unequal exposure of some social groups to "climate vulnerability"; as well as learning to develop the collective political capabilities needed to challenge these systems and transform them into something more sustainable and equitable (Chopin et al., 2018; Ryan, 2017; Waldron et al., 2019). As Fionnuala Waldron and her co-authors warn, "a focus on individual actions can deflect attention away from broader economic, social and political stakeholders and their responsibility to solve this crisis." In doing this, a whole "Green School" approach could end up, despite its other merits, promoting "an acritical form of education," through which "the status quo is upheld and 'business as usual' growth is justified" (Waldron et al., 2019, p. 898).

Adding to the Curriculum or Changing Paradigms?

If the global climate and environmental crises are being caused by human activity, and if these crises are having major impacts on human societies and economies, then our ability to understand the nature and significance of these crises, and develop collective strategies to address the causes and mitigate and adapt to the effects of these crises, requires far more than learning about climate science. It also demands that we draw on the social sciences and humanities to build an effective and empowering climate change education. "Climate change is … as much a societal problem as a physical one," declared the introductory editorial of the first issue of *Nature Climate Change* (Hefferman et al., 2011). "There's literally no way to think about climate change without becoming very quickly interdisciplinary," notes Bill McKibben (2016, pp. vii–viii) in his foreword to *Climate Change Across the Curriculum*, as "climate change is as much a problem for theologians, psychologists, poets, economists, political scientists, philosophers, anthropologists, and diplomats as it is for climatologists." Over the past decade, there has been a proliferation of work that explores the contributions of social science and humanities to understanding and acting on the climate crisis (e.g., Dunlap & Brulle, 2015; Elliott et al., 2017; Jorgenson et al., 2019). As challenging as it often is to get climate change addressed centrally and extensively in the science curriculum, this struggle remains deeply inadequate. For there is a need to ensure that climate change is addressed throughout the entire school and university curriculum as well (Fretz, 2015; Kavanagh et al., 2021; Kwauk, 2020). As David Orr (1994) has long argued, "all education is environmental education," not just science education.

It is possible, and probably likely, that even more than this is required to foster effective climate change education and action. Climate change researchers, theorists and activists point out that the root causes of global climate and environmental crisis lie not in simple technological and resource decisions that could be relatively straightforward to fix—e.g., replacing fossil fuels with renewables as our primary source of energy—but in fundamental problems with how contemporary societies and economies have been constructed, and how we have built our understanding of the social and physical worlds in which we live. If this is the case, then addressing this crisis requires a rethinking and reshaping of global cultural, social and economic systems. For climate change education to play an effective role in this process, it is no longer a matter of adding units to pre-existing school curriculum but shifting the paradigm(s) within which school learning and knowledge are developed. "We must immediately and collectively achieve a paradigm shift," argues UNESCO (2019, pp. 5, 8):

> [This] requires a change in mindset and a different way of looking at the relationship between humans and nature. ... 'Changing minds' means changing the way we think and refocusing our practices, our research, our technologies, our economies and our investments by developing a clear awareness of our responsibility towards our planet and its inhabitants. 'Changing minds'[is foundational] because it is in minds that virtuous circles of development must be first set in motion.

Others make parallel arguments. To tackle the climate crisis effectively, argues Christina Kwauk (2020, p. 12), we need to spark "a deep shift in consciousness about humanity's relationship with the more-than-human world," and "the existing inequities of our current human-to-human ... systems of relationships." For Selby and Kagawa (2018, p. 302), the aim of effective climate change education must be no less than "a stretching of epistemology so that unsustainable practices are challenged, taken-for-granted thinking and assumptions disrupted, root causes of global dysfunction interrogated, values subjected to critical scrutiny, change potential of socio-effective learning unleashed, and paradigm shift thus catalyzed."

Intellectually, there is already in existence an extraordinary richness of potential paradigm shifting frameworks from which climate change educators can draw. In environmental education, the new millennium has witnessed a flourishing of such models, including visions of ecological literacy (Stone & Barlow, 2005), ecopedagogy (Kahn, 2010; Misiaszek, 2020), transformative eco-education (Lin & Oxford, 2011), ecological intelligence (Goleman et al., 2012), ecojustice education (Martusewicz et al., 2014; Turner, 2015; Martusewicz, 2018), an eco-social approach to education (Lehtonen et al., 2019), and so on. Many focus on similar concerns. Ecojustice education, for example, which draws on the work of Chet Bowers (1993, 2001), begins with the assumption that "the ecological crisis is really a cultural crisis—that is, a crisis in the way people have learned to think and thus behave in relation to larger life systems and toward each other" (Martusewicz et al., 2014, p. 8). Ecojustice education seeks to help students "learn to think differently about our relationships to each other and to the natural world, and … to identify and revalue those critical practices of mutual support and interdependence that still exist in communities all over the world" (p. 8). This involves questioning the "hierarchized dualisms" at the center of "Western ways of thinking," that position men as superior to women, culture as superior to nature, reason as superior to emotion, and so forth (p. 57). It involves challenging the "logic of domination" that "underlies the acceptance and continuation of class inequality, along with gender and race inequalities, other forms of social degradation, *and* ecological devastation" (p. 63; emphasis in original). It also entails critiquing dominant "discourses of modernity," such as individualism, rationalism/scientism, commodification, consumerism, anthropocentrism, androcentrism, ethnocentrism, and the ideology of progress (pp. 66–67). The eco-social approach to education, as presented by Anna Lehtonen, Arto Salonen and Hannele Cantell (2019), proposes something much the same. "Combating climate change," argue Lehtonen et al. (2019, p. 341), "requires critical analysis of the reality we are constructing, and reflection on the roots of the problem including human-nature interdependency, individualism and consumerism." To do this will entail a "holistic change in thinking and action" and "widening the modern concept of knowledge" to challenge the "critical illusions of western culture" (p. 341). More specifically, it will involve "shifting from materialism towards post-materialism" (p. 343), "from segregated knowing and cultural dichotomies to holism and understanding interconnectedness" (p. 344), "from knowledge and rationalism to holism, intuition, embodied and emotional awareness" (p. 350), and "from individualism to creative, collective collaboration" (p. 353). All of this, in turn, requires fundamental changes in educational structure, curriculum and pedagogy (pp. 355–367).

Other educators have looked beyond environmental education to draw on theories and ideologies from other fields of practice that have been developed to rethink and rework core cultural, social and economic systems. Some seek to reimagine education in support of the social practice of commoning—of creating, reclaiming and protecting the commons, and participating in a broader commons movement (Bollier & Helfrich, 2019; Korsgaard, 2019; Means et al., 2017). Some seek to rethink education within a degrowth perspective, that rejects the embrace of endless growth long central to capitalist economies and societies (Getzin, 2019; Hickel, 2020; Jones, 2020; Kaufmann et al., 2019). Some look to the work of Ivan Illich and ideas of unlearning and unschooling as a basis for revisioning education to address social, economic, cultural as well as environmental crisis (Jain & Akomolafe, 2016; Neusiedl, 2021). In their proposal for "a new green learning agenda," Kwauk and Casey (2021) turn to feminist theory to develop a vision of "feminist transformative learning for climate action" (p. 63). The aim of such learning is to foster a "feminist planetary consciousness" the authors define as "an awareness of the interconnectivity of humanity's challenges and the state of the planet" and how "power and patriarchy, as well as colonialism and racism, impact both human and natural systems" (p. 4). This includes developing an understanding of "how an imbalance of power between men and women, the majority and the minority, settler colonialists and Indigenous groups, and so on, are just as destructive to life on this planet as an imbalance of greenhouse gases" (p. 55). It includes moving "away from a 'dominion over' worldview that positions the earth ... as resources to exploit for economic growth and development, and for power and control," and towards "a justice-oriented worldview of 'care for' and 'compassion toward' that strengthens the social bonds between people" and embraces an ethic of "environmental stewardship" (p. 55). It also involves fostering "interdisciplinary thinking, integrative thinking, and disruptive thinking" that can enable learners "to see what new socioecological structures and systems are possible, why they are desirable, and therefore why we should work collectively to achieve them" (pp. 59–60).

Finally, some educators look to Indigenous, traditional and non-Western frameworks for rethinking educational paradigms, based on the argument that if many of the central assumptions and approaches of Western knowledge appear to be at the heart of creating climate and environmental crisis, as well as other cultural, social, political and economic crises, there is value in looking to other frameworks of knowledge, learning and practice for guidance and inspiration. "Alternatives do exist" in many non-Western parts of the world, Boaventura de Sousa Santos and Maria Paula Meneses (2020, pp. xv, xvii) argue, for "the global South is now an extremely wide field of experiments in fighting for a better world." In some cases, the claim is that what differentiates Indigenous knowledge systems from dominant Western paradigms is a central concern with understanding and protecting the deep interconnectedness between human societies and the natural environment. Fikile Nxumalo and Libby Berg (2020, p. 46) speak of an "ethos of radical relationality" as being at the heart of Indigenous knowledges, which acts to "foreground an intrinsic interconnection between humans and more-than-human relatives, and … recognize the agentic sociality of land and waters and their inhabitants." In other cases, the claim is that Indigenous and non-Western frameworks (what Santos and Meneses (2020) call "epistemologies of the South") are defined by "ways of knowing" that were "born in struggles" of "a heterogeneous mass of subaltern groups," including "peasants and landless laborers, unemployed people, women, Indigenous and Afro-descendant peoples ... against the three modern forms of domination: capitalism, colonialism, and patriarchy" (pp. xiv–xvii). Educational theorists, researchers and practitioners are now trying to rethink educational practice and purpose based on what might be learned from particular Indigenous knowledge systems in different parts of the world (e.g., Nesterova, 2020; Vaioleti & Morrison, 2019); Indigenous knowledge systems more generally (e.g., Harmin et al., 2017; Smith et al., 2019); as well as newly emergent non-Western systems of knowledge and practice. In Latin America, for example, there is growing interest in the concept of *buen vivir* as an alternative model to dominant development paradigms.

Buen vivir is an "ethical paradigm," rooted in Indigenous knowledge systems, that focuses on "searching for harmony and conviviality between human beings and nature," while rejecting "the utilitarian relation of humans towards nature, the monocultural-colonial orientational pattern of state and the asymmetric structuring of society" (Weber & Tascón, 2020, p. 851). Education embracing *buen vivir* is organized around principles of "epistemological pluralism," "porosity of boundaries," "holism of learning," "cooperativism," "compassion and nonviolence," "collectivism," "meaningful livelihoods," and "living the present" (Brown & McCowan, 2018, pp. 320–321).

While a wealth of paradigm shifting ideas are there for the taking, the question of how any of these alternative frameworks might be used to fundamentally reorganize the overall structure, practice and purpose of schooling writ large is another matter. "We expect that educators, especially perhaps those working in formal educational contexts, may well find many of our proposals outlandish," admit Selby and Kagawa (2018, p. 317), at the end of an article in which they call for a radical overhaul of education to address the climate crisis. "Enacting transformative agendas can … be challenging in current educational contexts," Joseph Henderson and his colleagues concur (2017, p. 418). Instead, where these frameworks are being adopted at the moment is more locally: in Green Schools and alternative universities; in individual programs or classes within mainstream schools and universities; and in nonformal educational spaces that have been created by environmental and climate change activists outside of the formal education system. As with Green Schools, the hope is that these alternative, experimental educational settings can serve as vital prefigurative spaces for working through ideas and practices that could later be extended at scale. But the risk is that these programs end up providing excellent, specialist educational opportunities for learners who are already fully committed to doing everything possible to take action on the climate crisis, but who are then left frustrated why more people fail to share their passion and viewpoint. To avoid such an unwelcome scenario, it can be useful to consider other key elements of an effective, radical climate change education, beyond the matter of developing and selecting one or another of these alternative curricular paradigms: this includes consideration of place, self interest and power, certainly; but also issues of age, life stage and generation, of hope, fear and passion, and of how these formal and nonformal models of education could and should be linking up with wider fields of informal learning.

From Knowledge to Action? The Limitations of Curricular Movements for Change

In *Dust Bowls of Empire*, her historical analysis of the crisis of soil erosion ("dust-bowlification") that spread across the world in the first half of the twentieth century in the context of European and North American colonialist expansion, Hannah Holleman (2018) shows that this global environmental crisis didn't occur due to a lack of knowledge or understanding, but in spite of it. "Popular and scholarly accounts ... suggest the Dust Bowl was unforeseen and therefore unpreventable, but once it was recognized as a crisis, human perseverance and ingenuity resolved it," writes Holleman (p. 8). What the historical record actually shows, Holleman points out, is that by the late nineteenth century, "an extensive body of literature on soil erosion" existed that ensured that "soil erosion was a well-understood problem" for "policymakers and the reading public," including anyone "concerned with agriculture," from farmers to "economic, and land-owning elites" (pp. 78–79). In fact, Holleman notes, even as general knowledge and understanding of the causes of and solutions to soil erosion grew, the problem worsened. Holleman draws a direct link to the contemporary climate crisis:

> Greater scientific understanding did not prevent increased soil degradation any more than greater knowledge of climate science in more recent decades has prevented the quickening of climate change. Which is not to say that efforts to address erosion, such as changes to government policies and programs, have not made a difference, but that despite these improvements, the problem has expanded and worsened overall. (p. 78)

Holleman's conclusion is that "ecological crises are not resolved by increased scientific understanding, commonly held knowledge, [or] sophisticated technological development." Rather, they will only be solved if we can learn how to address "the social drivers of crises directly"—which in the case of soil erosion, at least, are to be found in the structural forces and cultural ideologies of capitalist and colonialist economic expansion (p. 73).

Christophe Bonneuil and Jean-Baptiste Fressoz make a similar argument in *The Shock of the Anthropocene*, their account of the emergence of the new geological era in which the impacts of human society have come to fundamentally reshape and damage the natural and physical environment. "The grand narrative of the Anthropocene," Bonneuil and Fressoz (2016, pp. 16–17, 167) note, is "the story of an awakening," in which, over a period of two centuries, human societies "unconsciously destroyed nature to the point of hijacking the Earth system into a new geological epoch," until finally, "in the late twentieth century, a handful of Earth system scientists finally opened our eyes," so that "now we are aware of the global consequences of human action." Unfortunately, the historical record does not support this claim. "The story of awakening is a fable," Bonneuil and Fressoz write, and "the opposition between a blind past and a clear-sighted present" is "historically false" (p. 17). In fact, "the period between 1770 and 1830 was marked ... by a very acute awareness of the interactions between nature and society," and "from the early nineteenth century onwards, we clearly see climate knowledge and discourse establishing ... connections" between human activity and climate change (pp. 172, 397). Like Holleman, Bonneuil and Fressoz make a case for needing to understand how climate change and environmental crisis grow worse despite widespread knowledge about these problems. The solution to environmental crises, the authors insist, lies in collective political action against:

> the institutions and oligarchies, the powerful symbolic and material systems, that led us into the Anthropocene: military apparatuses, the system of consumerist desire and its infrastructure, the gaps of income and wealth, the energy majors and the financial interests of globalization, the techno-scientific apparatuses when these work in commodity logics or silence criticisms and alternatives. (p. 650)

Historical accounts such as these help to show up one of the fundamental limitations of curriculum-focused movements to address the climate crisis. For these movements tend to rest on a belief that the greatest problem that has to be overcome to address this crisis is a lack of knowledge, understanding and belief among the general public and political and economic elites about the nature and significance of human induced climate change. As Kari Norgaard (2011, p. 64) notes, the "majority of research" on climate change communication and education has "presumed information was the limiting factor in public nonresponse," based on "the thinking that 'if people only knew the facts,' they would act differently." However, if widespread knowledge about environmental destruction can and regularly has co-existed with acceleration in the rate of this destruction, then clearly something more than enhanced curricular knowledge about the global environment and climate is essential.

What history helps to show is reinforced by research in sociology, psychology and communication studies: increased knowledge about climate and environmental crisis does not necessarily lead to action to address these crises effectively; and conversely, individuals and groups with relatively limited knowledge of the details of climate and environmental crises can be extraordinarily committed and effective in taking individual and collective action in attempts to address these crises. In the environmental education literature, one of the most well known articulations of this dilemma is Anja Kollmuss and Julian Agyeman's (2002) "Mind the Gap" article, which seeks to understand the gap that often exists "between the possession of environmental knowledge and environmental awareness, and displaying pro-environmental behavior" (p. 239). One of the most striking observations in this article is not the widely reported observation of the gap between having knowledge and concerns about environmental destruction and yet not engaging in substantive individual or collective action to address this destruction, but the converse finding. Reporting on earlier research by Kempton et al. (1995), Kollmuss and Agyeman note that some environmental activists may not have much greater knowledge about environmental issues than non- or anti-environmentalists, and the level of environmental knowledge among these activists can be quite low. While "people have to have a basic knowledge about environmental issues … in order to act pro-environmentally in a conscious way," Kollmuss and Agyeman argue, they do not necessarily need "very detailed technical knowledge" (p. 250). Along with other scholars, Kollmuss and Agyeman conclude that "only a small fraction of pro-environmental behavior can be directly linked to environmental knowledge and environmental awareness" (p. 250).

Unlike the historical analyses of Holleman, Bonneuil and Ferroz, which focus on the existence of a gap between knowledge and effective action at the level of society overall, the concern of environmental researchers like Kollmuss and Agyeman is the gap between knowledge and action at the level of the individual. But, much like the arguments made by the historians, Kollmuss and Agyeman (2002, pp. 248–256), and others who have followed them, effectively argue that this gap is moderated by cultural factors (ideologies, discourses or cultural logics that shape values, motivations, identities, emotions, attitudes and so on) as well as by social structural factors (the infrastructures, institutions and economies which shape and constrain the choices and actions that are made easily available to individuals and local communities). "Many pro-environmental behaviors can only take place if the necessary infrastructure is provided (e.g. recycling, taking public transportation)," Kollmuss and Agyeman note, and "the poorer such services are the less likely people are to use them" (p. 248). Over the last two decades, a significant body of literature has sought to address this problem of a knowledge-action gap, specifically as it relates to environmental and climate action. Much of this literature is based in psychology and communication studies and focuses on individual behavior change—some of which will be discussed in the following chapter (see, for example, Kenis & Mathijs, 2012; Kennedy et al., 2009; Moser & Dilling, 2011; Rousell & Cutter-MacKenzie-Knowles, 2020). But there are also a number of important sociologically oriented analyses (Kenis & Mathijs, 2012, p. 47; Norgaard, 2011). Kari Norgaard's (2011, p. 11) ethnography of a small community in Norway, for example, shows how "socially organized" processes of denial lead local community members to fail "to integrate [climate change] knowledge into everyday life or to transform it into social action." Or, to take a quite different kind of study, Simon Pirani's (2018) historical study of the massive expansion of fossil fuel consumption over the course of the twentieth century, shows how the development of global "technological, social and economic systems" strongly shape and limit individual choice and action in relation to their continuing use of fossil fuels as a primary source of energy.

In a summary of this recent literature, Anneleen Kenis and Erik Mathijs (2012, p. 58) conclude that "closing the 'gap' between knowledge and action does not in the first place—at least not in the case of environmentally aware citizens—require a further raising of people's knowledge of the environmental problem as such." Part of what is needed, the authors suggest, is a much broader definition of the kinds of knowledge and understanding that are vital for developing effective climate change action. Kenis and Mathjis talk of the importance of developing analyses of "the root causes of the environmental problem, visions of alternatives and a broad spectrum of strategies to realise them," which together can "enhance people's 'action competence'" and power to engage in effective "collective social action" (p. 47). To a degree, the whole school approaches and paradigm shifting conceptual frameworks discussed in this chapter constitute attempts to address these wider understandings of the kinds of knowledge needed to address the climate crisis. But there is also a need to recognize that effective climate change education requires more than just fostering new forms of knowledge, conceptual frameworks or belief systems, however useful, engaging and inspirational these might be. There are other pieces of the puzzle beyond the question of different types of curricular knowledge that are also essential. It is to these broader and alternative considerations that this book now turns. In so doing, it seeks to draw on the insights not just of those who work directly in the fields of climate change, environmental and sustainable education, but from others working in the older and wider traditions of radical education as well.

References

Albeck-Ripka, L. (2018). Idaho stripped climate change from school guidelines. Now, it's a battle. *New York Times*, 6 February. https://www.nytimes.com/2018/02/06/climate/idaho-schools-climate-change.html

Apple, M. (2013). Can education change society? Du Bois, Woodson and the politics of social transformation. *Review of Education, 1*(1), 32–56.

Apple, M. (2019). *Ideology and curriculum* (4th ed.). Routledge.

Apple, M., & Franklin, B. (2019). Curricular history and social control. In M. Apple (Ed.), *Ideology and curriculum* (4th ed., pp. 63–84). Routledge.

Au, W. (2012). *Critical curriculum studies: Education, consciousness, and the politics of knowing*. Routledge.

Bagley, K. (2015). Kids caught in crossfire of climate education battle. *Inside Climate News*, 29 January. https://insideclimatenews.org/news/29012015/kids-caught-crossfire-climate-education-battle/

Green School Bali. (2020). Admissions. https://www.greenschool.org/bali/admissions/

Bauman, Z. (2005). *Liquid life*. Polity Press.

BBC. (2013). Keep climate change lessons in curriculum, urge petitions. *BBC News*, 15 April. https://www.bbc.co.uk/news/education-22158941

Beard, M. (2020). Montessori education could reduce the advantage gap between rich and poor, but it's only available to the rich. *Guardian*, 19 January. https://www.theguardian.com/commentisfree/2020/jan/19/montessori-education-could-reduce-the-advantage-gap-between-rich-and-poor-but-its-only-available-to-the-rich

Bollier, D., & Helfrich, S. (2019). *Free, fair & alive: The insurgent power of the commons*. New Society.

Bonneuil, C., & Fressoz, J. (2016). *The shock of the Anthropocene*. Verso.

Bowers, C. (1993). *Education, cultural myths, and the ecological crisis: Toward deep changes*. State University of New York Press.

Bowers, C. (2001). *Educating for eco-justice and community*. University of Georgia Press.

Branch, G., Rosenau, J., & Berbeco, M. (2016). Climate education in the classroom: Cloudy with a chance of confusion. *Bulletin of the Atomic Scientists* 72(2): 89–96. New York: Routledge.

Brown, E., & McCowan, T. (2018). *Buen vivir*: Reimagining education and shifting paradigms. *Compare, 48*(2), 317–323.

Chertoff, E. (2013). Reggio Emilia: From postwar Italy to NYC's toniest preschools. *Atlantica*, 17 January. https://www.theatlantic.com/national/archive/2013/01/reggio-emilia-from-postwar-italy-to-nycs-toniest-preschools/267204/

Chopin, N., Hargis, K., & McKenzie, M. (2018). *Building climate-ready schools: Towards identifying good practices in climate change education*. Sustainability and Education Policy Network.

Climate Scorecard. (2020). *Global spotlight report #23: Climate change education*. https://www.climatescorecard.org/project/global-spotlight-report-23/

Colston, N., & Ivey, T. (2015). (Un)Doing the next generation science standards: Climate change education actor-networks in Oklahoma. *Journal of Education Policy, 30*(6), 773–795.

Colston, N., & Vadjunec, J. (2015). A critical political ecology of consensus: On 'Teaching Both Sides' of climate change controversies. *Geoforum, 65*, 255–265.

Cooper, S. (2018). In America's science classrooms, the creep of climate skepticism. *Undark*. 2 November. https://undark.org/2018/11/02/climate-change-science-textbooks-classrooms/

Coughlan, S. (2017). Did Michael Gove really try to stop teaching climate change? *BBC News*. 12 June. https://www.bbc.co.uk/news/education-40250214

Cuban, L. (1993). The lure of curricular reform and its pitiful history. *Phi Delta Kappan, 75*(2), 182–185.

Czajka, K. (2019). States are introducing bills that could prevent teachers from advocating for climate change. *Pacific Standard*, 18 February. https://psmag.com/news/state-bills-could-prevent-teachers-from-advocating-for-climate-change

Dawson, A. (2017). *Extreme cities*. Verso.

Dawson, J. (2020). Idaho lawmaker: Teaching kids about climate change imparts 'Greta Thunberg Syndrome.' *Boise State Public Radio*, 23 January. https://www.boisestatepublicradio.org/post/idaho-lawmaker-teaching-kids-about-climate-change-imparts-greta-thunberg-syndrome#stream/0

Donnelly, K. (2020). How is only teaching climate catastrophism in schools discussion and debate? *The Spectator*, 26 February. https://www.spectator.com.au/2020/02/how-is-only-teaching-climate-catastrophism-in-schools-discussion-and-debate/

Drewes, A., Henderson, J., & Mouza, C. (2018). Professional development design considerations in climate change education: Teacher enactment and student learning. *International Journal of Science Education, 40*(1), 67–89.

Dunlap, R., & Brulle, R. (Eds.). (2015). *Climate change and society: Sociological perspectives*. Oxford University Press.

Eaton, E., & Day, N. (2020). Petro-pedagogy: Fossil fuel interests and the obstruction of climate justice in public education. *Environmental Education Research, 26*(4), 457–473.

Elliott, A., Cullis, J., & Damodaran, V. (Eds.). (2017). *Climate change and the humanities*. Palgrave Macmillan.

Energy Mix. (2020). Alberta aims for different 'balance' in climate change curriculum. *Energy Mix*, 31 January. https://theenergymix.com/2020/01/31/alberta-aims-for-different-balance-in-climate-change-curriculum/

Firozi, P. (2019). The energy 202: A wave of state bills could threaten science and climate education. *Washington Post*, 11 March. https://www.washingtonpost.com/news/powerpost/paloma/the-energy-202/2019/03/11/the-energy-202-a-wave-of-state-bills-could-threaten-science-and-climate-education/5c8569bc1b326b2d177d603f/

Foran, C. (2014). The plan to get climate change denial into schools. *Atlantica*, 8 December. https://www.theatlantic.com/education/archive/2014/12/the-plan-to-get-climate-change-denial-into-schools/383540/

Fretz, E. (Ed.). (2015). *Climate change across the curriculum*. Lexington Books.

Getzin, S. (2019). *Shifting education towards sustainability: How degrowth can transform education for sustainable development*. University of Zurich.

Givens, J. (2021). *Fugitive pedagogy: Carter G. Woodon and the art of black teaching*. Harvard University Press.

Goleman, D., Bennett, L., & Barlow, Z. (2012). *Ecoliterate: How educators are cultivating emotional, social and ecological intelligence*. Jossey-Bass.

Gorlewski, J., & Nuñez, I. (2020). Activism and social movement building in curriculum. Oxford research encyclopedia of education. https://doi.org/10.1093/acrefore/9780190264093.013.1421

Gough, A. (2020a). Educating Australia on the climate crisis. *Policy Forum*, 5 February. https://www.policyforum.net/educating-australia-on-the-climate-crisis/

Gough, A. (2020b). Transforming education through green schools: Trials, tribulations and tensions. In A. Gough, J. Lee, & E. Tsang (Eds.), *Green schools globally: Stories of impact on education for sustainable development* (pp. 412–438). Springer.

Gough, A., Lee, J., & Tsang, E. (Eds.). (2020). *Green schools globally: Stories of impact on education for sustainable development*. Springer.

Gruenewald, D., & Manteaw, B. (2007). Oil and water still: How No Child Left Behind limits and distorts environmental education in US schools. *Environmental Education Research, 13*(2), 171–188.

Hargis, K., Chopin, N., & McKenzie, M. (2018). *Ten Canadian schools' stories of climate action*. Sustainability and Education Policy Network.

Harmin, M., Barrett, M., & Hoessler, C. (2017). Stretching the boundaries of transformative sustainability learning: On the importance of decolonizing ways of knowing and relations with the more-than-human. *Environmental Education Research, 23*(10), 1489–1500.

Hefferman, O., Thorpe, J., & Brown, A. (2011). Whole-system science. *Nature Climate Change, 1*(1), 1.

Henderson, J., & Drewes, A. (2020). Teaching climate change in the United States. In J. Henderson & A. Drewes (Eds.), *Teaching climate change in the United States*. Routledge.

Henderson, J., et al. (2017). Expanding the foundation: Climate change and opportunities for educational research. *Educational Studies, 53*(4), 412–425.

Hickel, J. (2020). *Less is more: How degrowth will save the world.* William Heinemann.

Holland, C. (2020). The implementation of the Next Generation Science Standards and the tumultuous fight to implement climate change awareness in science curricula. *Brock Education Journal, 29*(1), 35–52.

Holleman, H. (2018). *Dust bowls of empire: Imperialism, environmental politics, and the injustice of 'green' capitalism.* Yale University Press.

Huckle, J. (2013). Eco-schooling and sustainability citizenship: Exploring issues raised by corporate sponsorship. *Curriculum Journal, 24*(2), 206–223.

Hursh, D., Henderson, J., & Greenwood, D. (2015). Environmental education in a neoliberal climate. *Environmental Education Research, 21*(3), 299–318.

Ibrahim, Z. (2020). Teach the Future unveils Climate Emergency Education Bill. *Students Organising for Sustainability UK* 28 February. https://sustainability.nus.org.uk/articles/teach-the-future-unveils-climate-emergency-education-bill

Jain, M., & Akomolafe, B. (2016). This revolution will not be schooled. In A. Skinner et al. (Eds.), *Education, learning and the transformation of development.* Routledge.

Jones, A. (2020). What is an educational good? Theorising education as degrowth. *Journal of Philosophy of Education.* https://doi.org/10.1111/1467-9752.12494

Jorgenson, A., et al. (2019). Social science perspectives on drivers of and responses to global climate change. *WIREs Climate Change, 10*(1), e554.

Kahn, R. (2010). *Critical pedagogy, ecoliteracy & planetary crisis: The ecopedagogy movement.* Peter Lang.

Kaufman, M. (2020). The carbon footprint sham. *Mashable.* https://mashable.com/feature/carbon-footprint-pr-campaign-sham/?europe=true

Kaufmann, N., Sanders, C., & Wortmann, J. (2019). Building new foundations: The future of education from a degrowth perspective. *Sustainability Science, 14*, 931–941.

Kavanagh, A., Waldron, F., & Mallon, B. (2021). *Teaching for social justice and sustainable development across the primary curriculum*. Routledge.

Kempton, W., et al. (1995). *Environmental values in American culture*. MIT Press.

Kenis, A., & Mathijs, E. (2012). Beyond individual behaviour change: The role of power, knowledge and strategy in tackling climate change. *Environmental Education Research, 18*(1), 45–65.

Kennedy, E., et al. (2009). Why we don't 'walk the talk': Understanding the environmental values/behavior gap in Canada. *Human Ecology Review, 16*(2), 151–160.

King, L., Crowley, R., & Brown, A. (2010). The forgotten legacy of Carter G. Woodson: Contributions to multicultural social studies and African American history. *The Social Studies, 101*, 211–215.

Kollmuss, A., & Agyeman, J. (2002). Mind the gap: Why do people act environmentally and what are the barriers to pro-environmental behavior? *Environmental Education Research, 8*(2), 239–260.

Korsgaard, M. (2019). Education and the concept of commons: A pedagogical reinterpretation. *Educational Philosophy and Theory, 51*(4), 445–455.

Koziol, M. (2020). 'We want to see balance': Right-wing activists target primary school children on climate. *Sydney Morning Herald*, 16 February. https://www.smh.com.au/national/we-want-to-see-balance-right-wing-activists-target-primary-school-children-on-climate-20200214-p540ti.html

Kwauk, C. (2020). *Roadblocks to quality education in a time of climate change*. Brookings.

Kwauk, C., & Casey, O. (2021). *A new green learning agenda: Approaches to quality education for climate action*. Brookings.

Labaree, D. (2008). The winning ways of a losing strategy: Educationalizing social problems in the United States. *Educational Theory, 58*(4), 447–460.

Lawrynuik, S. (2019). 'It's kind of frightening': Students worry climate change education lacking in Alberta classrooms. *The Narwhal*, 10 July. https://thenarwhal.ca/its-kind-of-frightening-students-worry-climate-change-education-lacking-in-alberta-classrooms/

Lehtonen, A., Salonen, A., & Cantell, H. (2019). Climate change education: A new approach for a world of wicked problems. In J. Cook (Ed.), *Sustainability, human well-being, and the future of education* (pp. 339–374). Palgrave Macmillan.

Lin, J., & Oxford, R. (Eds.). (2011). *Transformative eco-education for human and planetary survival.* IAP.
Mann, M. (2021). *The new climate war.* PublicAffairs.
Mann, M., & Toles, T. (2016). *The madhouse effect.* Columbia University Press.
Martusewicz, R. (2018). *A pedagogy of responsibility.* Routledge.
Martusewicz, R., Edmundson, J., & Lupinacci, J. (2014). *Ecojustice education: Toward diverse, democratic, and sustainable communities.* Routledge.
McKibben, B. (2016). Foreword. In E. Fretz (Ed.), *Climate change across the curriculum* (pp. vii–viii). Lanham: Lexington Books.
Means, A., Ford, D., & Slater, G. (2017). *Educational commons in theory and practice.* Palgrave Macmillan.
Melia, M. (2019). Climate change skeptics target school curriculum to combat 'global warming propaganda.' *Chicago Tribune*, 6 March. https://www.chicagotribune.com/news/environment/ct-climate-change-school-curriculum-20190306-story.html
Misiaszek, G. (2020). *Ecopedagogy: Critical environmental teaching for planetary justice and global sustainable development.* Bloomsbury.
Monroe, M., et al. (2019). Identifying effective climate change education strategies: A systematic review of the research. *Environmental Education Research, 25*(6), 791–812.
Moser, S., & Dilling, L. (2011). Communicating climate change: Closing the science-action gap. In J. Dryzek, R. Norgaard, & D. Schlosberg (Eds.), *The Oxford handbook of climate change and society* (pp. 161–174). Oxford University Press.
NCSE—National Center for Science Education. (2016). *Mixed messages: How climate change is taught in America's public schools.* NCSE.
NCSE & TFN—National Center for Science Education & Texas Freedom Network. (2020). *Making the grade? How state public school science standards address climate change.* NCSE.
Nesterova, Y. (2020). Rethinking environmental education with the help of indigenous ways of knowing and traditional ecological knowledge. *Journal of Philosophy of Education, 54*(5), 1047–1052.
Neusiedl, C. (2021). *Revolutions in learning and education from India.* Routledge.
Norgaard, K. (2011). *Living in denial: Climate change, emotions, and everyday life.* MIT Press.
NSTA—National Science Teachers Association. (2018). The teaching of climate science. https://www.nsta.org/nstas-official-positions/teaching-climate-science

Nxumalo, F., & Berg, L. (2020). Conversations on climate change pedagogies in a Central Texas kindergarten classroom. In J. Henderson & A. Drewes (Eds.), *Teaching climate change in the United States*. Routledge.

Orr, D. (1994). *Earth in mind*. Island Press.

Otto, E. (2017). Evidence for the 'slippery slope' to censorship: The story from Florida and Collier county. *Journal of Intellectual Freedom & Privacy, 2*(3), 20–30.

Pirani, S. (2018). *Burning up: A global history of fossil fuel consumption*. Pluto.

Reid, A. (2019). Climate change education and research: Possibilities and potentials versus problems and perils? *Environmental Education Research, 25*(6), 767–790.

Rousell, D., & Cutter-MacKenzie-Knowles, A. (2020). A systematic review of climate change education: Giving children and young people a 'voice' and a 'hand' in redressing climate change. *Children's Geographies, 18*(2), 191–208.

Ryan, E. (2017). *Localising the global eco-schools program in South Africa: A postcolonial analysis*. PhD dissertation, Southern Cross University.

Santos, B., & Meneses, M. (Eds.). (2020). *Knowledges born in the struggle: Constructing the epistemologies of the Global South*. Routledge.

Selby, D., & Kagawa, F. (2018). Teetering on the brink: Subversive and restorative learning in times of climate turmoil and disaster. *Journal of Transformative Education, 16*(4), 302–322.

Smith, L., Tuck, E., & Yang, K. (Eds.). (2019). *Indigenous and decolonizing studies in education*. Routledge.

Stevenson, R. (1987/2007). Schooling and environmental education: Contradictions in purpose and practice. *Environmental Education Research, 13* (2): 139–153.

Stone, M., & Barlow, Z. (2005). *Ecological literacy: Educating our children for a sustainable world*. Sierra Club Books.

Teach the Future. (2020a). Asks. https://www.teachthefuture.uk/hub/b978de45-e275-4fa9-981c-fd4ee7ee61a5

Teach the Future. (2020b). Climate Emergency Education Bill. https://www.teachthefuture.uk/hub/b978de45-e275-4fa9-981c-fd4ee7ee61a5

Teach the Future. (2020c). Our case for investing in climate education. https://www.teachthefuture.uk/hub/b978de45-e275-4fa9-981c-fd4ee7ee61a5

Teach the Future. (2020d). Teach the Future. https://www.teachthefuture.uk

Teach the Future. (2020e). Teach the Future England. https://www.teachthefuture.uk/england

TES—Times Education Supplement. (2020). Law proposed to tackle climate change fears in schools. *Times Education Supplement*, 10 February. https://www.tes.com/news/law-proposed-tackle-climate-change-fears-schools

Tickle, L. (2020). Pupils draft their own climate bill as anxiety grows over lack of guidance for schools. *Guardian*, 11 February. https://www.theguardian.com/education/2020/feb/11/pupils-draft-climate-bill-anxiety-lack-guidance-schools

TNT—Truth in Textbooks. (2020). FAQ. Truth in textbooks. https://truthintexastextbooks.com/faq/

Turner, R. (2015). *Teaching for EcoJustice*. Routledge.

Turner, C. (2020). Climate change activists are 'commandeering' school curriculum, Ofsted chief warns. *Telegraph*, 1. December 1.

Tyack, D., & Tobin, W. (1994). The 'grammar' of schooling: Why has it been so hard to change? *American Educational Research Journal, 31*(3), 453–479.

UNESCO. (2016). *Getting climate-ready: A guide for schools on climate action*. UNESCO.

UNESCO. (2019). *Changing minds, not the climate!* UNESCO.

Vaioleti, T., & Morrison, S. (2019). The value of indigenous knowledge to education for sustainable development and climate change education in the Pacific. In E. McKinley & L. Smith (Eds.), *Handbook of indigenous education* (pp. 651–670). Springer.

Vanderbilt, T. (2018). Welcome to the jungle: The Bali school attracting wealthy western families. *Sydney Morning Herald*, 2 February. https://www.smh.com.au/lifestyle/welcome-to-the-jungle-the-bali-school-attracting-wealthy-western-families-20180117-h0jri9.html

Waldron, F., Ruane, B., Oberman, R., & Morris, S. (2019). Geographical process or global injustice? Contrasting educational perspectives on climate change. *Environmental Education Research, 25*(6), 895–911.

Wals, A., & Benavot, A. (2017). Can we meet the sustainability challenges? The role of education and lifelong learning. *European Journal of Education, 52*, 404–413.

Weber, S., & Tascón, M. (2020). Pachamama-La Universidad del 'Buen Vivir': A First Nations sustainability university in Latin America. In W. Filho et al. (Eds.), *Universities as living labs for sustainable development* (pp. 849–862). Springer.

Weyrauch, E. (2017). Florida's textbooks are a new battleground in America's fight over facts. *Time*, 30 August. https://time.com/4910125/textbook-law-florida/

Woodson, C. (1933). *The mis-education of the Negro*. CreateSpace.

Worth, K. (2017). A new wave of bills takes aim at science in the classroom. *Frontline*, 8 May. https://www.pbs.org/wgbh/frontline/article/a-new-wave-of-bills-takes-aim-at-science-in-the-classroom/

Worth, K. (2018a). Dueling books compete to educate kids on climate change. *Frontline*, 2 November. https://www.pbs.org/wgbh/frontline/article/dueling-books-compete-to-educate-kids-on-climate-change/

Worth, K. (2018b). Mailings to teachers highlight a political fight over climate change in the classroom. *Frontline*, 23 March. https://www.pbs.org/wgbh/frontline/article/mailings-to-teachers-highlight-a-political-fight-over-climate-change-in-the-classroom/

Worth, K. (2019). Inside Idaho's long legislative battle over climate change education. *Frontline*, 20 December. https://www.pbs.org/wgbh/frontline/article/inside-idahos-long-legislative-battle-over-climate-change-education/

Yoder, K. (2020). Footprint fantasy. *Grist*, 26 August. https://grist.org/energy/footprint-fantasy/

Zabjek, A. (2020). In Alberta classrooms, oil is a tricky lesson. *The Sprawl*, 7 May. https://www.sprawlcalgary.com/climate-change-curriculum-alberta

3

Nudging and Informal Learning

The claims made by the Virginia-based Center for Behavior and the Environment about the power of nudges to address global climate crisis are not modest. More than one third of greenhouse gas emissions could be eliminated worldwide over the next thirty years by using nudges and related individual behavior change mechanisms, claims the center—an organization that was set up to promote social marketing techniques and behavioral science insights to address global environmental problems (Williamson et al., 2018). "When it comes to a problem as immense as climate change," journalist Emily Bobrow (2018) writes, "it is easy to feel discouraged by the seemingly piddling effects of personal behavior." "A few effective nudges here and there," Bobrow reflects, "hardly seem sufficient in the absence of grand changes in environmental policy." But the Center for Behavior and the Environment disagrees. "Individual behavior change," Center for Behavior researchers insist, "when taken up by billions of people makes a decisive difference," and can produce "a transformative contribution to solving the climate crisis" (Williamson et al., 2018, p. 5).

Nor is the Center for Behavior and Environment alone in making such strong claims. Over the last decade, a growing number of elite

S. Tannock, *Educating for Radical Social Transformation in the Climate Crisis*, Palgrave Studies in Education and the Environment,
https://doi.org/10.1007/978-3-030-83000-7_3

organizations have signed up to the power of the nudge—and individual behavioral change more generally—for addressing the climate crisis. In 2015, the World Bank titled its annual world development report, *Mind, Society and Behavior*, focusing on the possibilities that behavioral science offers for addressing a range of development problems, including climate change; the Bank also set up the Global Insights Initiative, sometimes referred to as its "global nudge unit," to incorporate insights of behavioral science into development projects (Rutter, 2016; World Bank, 2015). In 2016, the United Nations followed suit, appointing its first ever Behavioral Science Advisor and launching the UN Behavioral Initiative. For the UN, little nudges—"small, subtle, and sometimes counter-intuitive changes to the way a message or choice is framed, or how a process is structured"—through reshaping individual behavior, can have an "out-sized impact" even on planetary level problems such as climate change (Shankar & Foster, 2016; UN, 2017). In 2017, the OECD also turned its attention to what it called "behavioral insights," finding that the lessons of behavioral science were being drawn on by "nudge units" in close to two dozen countries, again to address a range of policy concerns, including climate change (OECD, 2017a, 2017b).

These organizations often work together to promote nudging in climate change practice. In 2018, Rare—the conservation organization behind the Center for Behavior and Environment—along with the United Nations Development Program, Nature Conservancy, World Wildlife Fund, National Geographic, and Conservation International co-sponsored a global crowdsourcing competition called *Solution Search: Climate Change Needs Behavior Change,* to "identify, spotlight, and reward nonprofits, governments, companies, or other organizations that are changing consumer habits using behavioral science principles." Finalists in the competition were invited to a workshop in Washington, DC, that was led by "behavioral experts" from the UK's Behavioural Insights Team, a group that was one of the first and perhaps most influential of the world's rapidly proliferating set of government and organization-based nudge units (Solution Search, 2018). The Grand Prize winners ended up being a cosmetics entrepreneur from Australia who developed a mobile phone App that nudges privileged and affluent women and girls to adopt greener consumer choices; and an American

NGO called Solar Sister that "taps rural African women's trust networks" to recruit friends, family and neighbors in low income countries in Africa to purchase branded, solar-powered products sold by the NGO (Solution Search, 2020).

The remarkable worldwide spread of elite interest in nudging and other individual behavior change mechanisms as a key way to tackle climate change (along with other social problems), has led some to talk of environmental behavior change governance as now being "an assumed orthodoxy amongst the majority of policy makers in both developed and developing world nations" (Barr, 2018, p. 749). Others write of the rise of the "psychological state" (Jones et al., 2013), "behavioural state" (Feitsma, 2018) or "behavioural change state" (Leggett, 2014). From the point of view of education, nudging holds particular interest as it claims to address one of the most intractable challenges that has been identified by environmental and social justice education. This is the gap that often exists between knowledge and action; or, as Kollmuss and Agyeman (2002, p. 240) put it, "the gap between the possession of environmental knowledge and environmental awareness, and displaying pro-environmental behavior." "Despite knowing that we need to adapt our behaviours to combat climate change, we may still not take the actions needed in our daily lives to be impactful," the United Nations Environmental Programme (UN, 2017) points out: nudging (allegedly) can help us close "the chasm … between good intentions and actual actions."

Nudging also holds interest for the field of education for less positive reasons. The prevailing policy focus on nudging and individual behavior change mechanisms often seems to displace education from the picture of effective climate change responses altogether. For example, though the IPCC's (2018) *Global Warming of 1.5 °C* report emphasizes "education and learning" as being central to addressing the climate crisis (p. 337), virtually all of the report's citations on this topic refer to work done in the behavioral sciences. Nudging has also been attacked by critics as being *un-educational* or *anti-educational*. Extensive use of nudging, some warn, can "impair … the learning capacity of modern societies," and "paves the way towards a society that has entirely unlearned to actively transform itself" (Bornemann & Burger, p. 216). "The true alternative to nudging," argues Gigerenzer (2015, p. 375), "is education:" and it is education, *not* nudging, according to critics like Gigerenzer that holds real promise for projects of radical, progressive social change.

There is now an enormous literature on nudging in general, as well as nudging done specifically in relation to the climate crisis and other environmental concerns—"green nudges." This chapter reviews some of this literature on nudging, green nudges and the climate crisis, while making links with the field of education studies. To incorporate nudging into a broader theory of education for radical social change, and retain the insights of the nudging approach while challenging its more problematic elements, the chapter argues that it is useful to link nudging with two well established educational traditions that have largely been missing from critical discussions of nudging: the critical sociology of education and its concept of the hidden curriculum; and the progressive education tradition, that promotes and makes use of a wide range of holistic, experiential and practical approaches to student learning.

Green Nudges and the Climate Crisis

In Copenhagen, at the Danish Executive Summit, where more than 500 Scandinavian CEOs gather together each year, researchers reduced the diameter of buffet dinner plates by a mere three centimeters, and found they were able to reduce the food waste produced by these distinguished business leaders by more than a quarter (Hansen, 2013). A couple of hundred miles north, in Gothenburg, Sweden, simply by redesigning a restaurant's menu to highlight vegetarian rather than meat-based meals, local citizens dining out were found to substantively increase their uptake of vegetarian dishes, and thereby reduce greenhouse gas emissions linked with their restaurant habits by more than half (Gravert & Kurz, 2021). In New Jersey, Rutgers University made a small change to its printing policy that enabled it to reduce its paper consumption by 44% over a four year period, saving 55 million sheets of paper, or "the equivalent of 4,650 trees": it switched the default option on campus printers from single sided to double sided printing (Sunstein & Reisch, 2014, p. 133). In Germany, a utilities company was similarly able to increase customer uptake of green electricity by changing its default power supply option from grey (i.e., fossil fuel and nuclear) to green energy (Pichert & Katsikopoulos, 2008). In Malmö, city officials launched a "No Ridiculous Car Trips" multimedia campaign, that used humor, visual publicity and citizen participation to draw "attention to the often habitual choice to drive short distances" in the city and foster a "public commitment to changing behavioural norms." It did this by inviting residents "to submit written accounts about their most nonsensical car trips for a chance to win bicycles," giving "small gifts to cyclists … as thanks for choosing to bike," and asking cyclists wearing distinctive bright color clothing "to time specific routes around the city to prove the convenience and quickness of cycling." Within a year, Malmö reported a 50% increase in cycling among residents, and significant drop in the number of short trips made in cars (UNEP, 2017, pp. 34–35; see also Lenhart, 2013). All of these are examples of green nudges: interventions that aim to encourage people to take up environmentally friendly practices, but without forcing or financially incentivizing them to do so (Schubert, 2017).

Introduced in a 2008 book, *Nudge*, by Richard Thaler and Cass Sunstein, the concept of a "nudge" is used by the authors to refer to any intervention "that alters people's behavior in a predictable way without forbidding any options or significantly changing their economic incentives" (p. 6). Though the terms appear rarely in their book, Thaler and Sunstein's work on nudging is essentially based on a theory of education and learning, and how we tend to learn to do things in the world in particular ways, and sometimes, change the way we are used to doing these things. More specifically, nudging is concerned with a type of informal learning called "tacit" or "unintentional" learning that is "unplanned and unconscious," and involves the adoption and "internalization of values, attitudes, behaviors, skills, etc., that occur unconsciously during everyday life" (Schugurensky, 2006, p. 167). The argument that Thaler and Sunstein make about nudging is that much of the most important and powerful learning we do—for example, learning to take effective action to address the problem of climate crisis—is not through formal or non-formal education, in which we are presented with an explicit curriculum that marshals together theory, evidence, argument and analysis to teach us something in a deliberate, conscious and reflective manner. Rather, it happens through informal learning, that is often unconscious, uncontrolled and unintentional (the authors use the term "Automatic System" to refer to such informal learning (pp. 19–20)). Such informal learning tends to be holistic and draws on past experience, repeat practice, social interaction with others, and an embodied and emotional engagement with the world. It is through informal learning, Thaler and Sunstein argue, that we learn to speak fluently in one or more languages, and that skilled practitioners develop their finely honed ability "to size up complex situations rapidly and to respond with both amazing accuracy and exceptional speed" (pp. 20–21). In making this argument, Thaler and Sunstein are recapping something which has long been pointed out by scholars of informal learning. Schugurensky (2006, p. 163), for example, notes that "much of the relevant (in the sense of personally meaningful and significant) learning acquired throughout our lives occurs in the area of informal learning." This includes, Schugurensky suggests, "the area of political and civic learning" that any effective climate change education program would need to address (p. 163).

The second core argument that Thaler and Sunstein make is that informal learning is massively influenced by local, everyday environmental contexts—and here, context includes physical, social, verbal, textual and visual context. "Small and apparently insignificant details can have major impacts on people's behavior," the authors write: "A good rule of thumb is to assume that 'everything matters'" (p. 3). The term used by Thaler and Sunstein to identify this environmental impact on informal learning is "choice architecture," which refers to how "the context in which people make decisions" has been organized. Choice architecture is shaped by human activity, but the lessons taught by this architecture may be "unintentional" (Thaler & Sunstein, 2008, p. 10): they may have been created accidentally, by chance, or alternatively, through the indirect influence of prevailing social discourses, cultural patterns and preferences. Sometimes, however, choice architecture can be intentionally designed to promote the learning or adoption of preferred values, principles, practices, habits and identities. The theory and practice of nudging aims to intentionally use choice architecture design—for example, changing the size of dinner plates at an executive banquet—to reshape the informal learning that people do, even if this informal learning is often unintentional and unconscious on the part of the learners. Again, though Thaler and Sunstein don't refer to the concept of informal learning, this theory and practice has long been recognized by the field of informal learning research. "Some unintentional learning is in fact intended—but not by the learner," writes Rogers (2014, pp. 18–19): "The many advertisements we are subjected to every day, the campaigns against smoking, unhealthy eating, HIV/AIDS etc are examples of learning that is not intended by the learner but is intended by the learning-provider."

Beyond these two core ideas (the importance of informal/tacit learning and local environmental contexts in shaping this learning), Thaler and Sunstein's book introduces two further key ideas. *Nudge* is a product of the neoliberal era in which it was written, characterized by a hostility toward direct state intervention in society and the economy to support collective, public welfare agendas, and a tendency to hold individuals responsible for their own well-being as well as the overall state of society, through the individual choices they make and actions they undertake. In Thaler and Sunstein's account, nudging is introduced not as an important kind of social education or intervention to consider *in addition* to other types of education or intervention: it is promoted as a cheaper, easier, more effective and more desirable *alternative* to stronger, more direct kinds of education and intervention. "We favor nudges over commands, requirements, and prohibitions," write Thalter and Sunstein (p. 10). Nudges are preferred because they "cost little or nothing," "impose no burden on taxpayers," and address social problems "in politically more palatable ways" (pp. 13, 189). "The sheer complexity of modern life," claim the authors, "undermine[s] arguments for rigid mandates" (p. 253). Instead, they argue "that better governance requires less in the way of government coercion and constraint, and more in the way of freedom to choose": ideally, government should be "smaller and more modest" (p. 14). Furthermore, in their discussion of how to improve a wide range of issues related to "health, wealth and happiness," Thaler and Sunstein focus on how to use nudges to reshape everyday decisions made by individuals. They do not address how we need to use social education and intervention to reshape the practices of the state, corporations or elites to improve our collective health and well-being. Nudging, in this sense, becomes part of what Maniates (2001, p. 43) critiques as the "individualization of responsibility, that "shifts blame [for environmental and other social problems] from State elites and powerful producer groups to more amorphous culprits like 'human nature' or 'all of us.'"

Nudge was also born in an age dominated by what Shosanna Zuboff (2019) calls "surveillance capitalism," characterized by increasingly extensive and intrusive practices of surveilling and manipulating human behavior. One core aspect of surveillance capitalism, Zuboff argues, beyond the manipulation of human behavior by corporations such as Google, Facebook and Microsoft for profit, is the rise of "unprecedented asymmetries in knowledge and the power that accrues to knowledge" between a technocratic elite, who surveil and manipulate, and the mass of humanity, who become fodder for elite surveillance and manipulation (pp. 17). These asymmetries, warns Zuboff, threaten to create a:

> division of learning … [that] fixes us in a new inequality marked by the tuners and the tuned, the herders and herded, the raw material and its miners, the experimenters and their unwitting subjects, those who will the future and those who are shunted toward others' guaranteed outcomes. (p. 317)

This embrace of knowledge asymmetry also marks Thaler and Sunstein's work on nudging: indeed, the authors favor what they call "asymmetric paternalism" as a desirable model for nudging (p. 249). Though Thaler and Sunstein acknowledge the remarkable power and effectiveness of tacit learning (in their words, the "Automatic System"), much of their book is given over to disparaging such learning as evidence of human fallibility and cognitive deficit. Most people thus tend to be "mindless, passive decision makers," who end up making "pretty bad decisions" (pp. 5, 37). We are prone to "self-control problems," often succumbing to "temptation," and "following the herd" (pp. 40, 44, 53). Our ideas and actions are irrational, "flawed and biased," and filled with elementary "blunders" (pp. 7, 17). The authors are particularly concerned with those in society they label as the "least sophisticated," who are liable to make the worst decisions of any of us (p. 249). By contrast, Thaler and Sunstein are enamored with the promise of an educated and empowered scientific elite, who can be trusted to shepherd this flawed mass of humanity into an improved state of collective well being. They call for "self-conscious efforts, by institutions in the private sector and also by government, to steer people's choices in directions that will improve their lives," efforts that should be based upon "the emerging science of choice, consisting of careful research by social scientists over the past four decades" (pp. 5, 7).

This combination of ideas—a grounded, pragmatic, real world theory of learning and action wedded to a neoliberal, technocratic model of elite intervention and management—helped lead to the extraordinary popularity of nudging among elite institutions over the last decade (Selinger & Whyte, 2012). In nudging, policy makers could find a way to tackle (or be seen to be tackling) enormous problems like climate crisis without any great cost or political struggle, since "it support[ed] a shift toward encouraging individual responsibility and a reduced reliance on regulations and laws to enforce change" (Barr, 2018, p. 748). Such popularity led to a proliferation of research and experiment with nudges in all policy areas, including green nudges to address climate crisis and other environmental problems. New organizations were launched that were dedicated to exploring the possibilities of green nudges: the Center for Behavior and the Environment in the United States, GreeNudge in Norway, and iNudgeYou in Denmark. Key types of nudges were identified that could be used to foster positive climate change action: changing the physical environment, shifting default options, reframing textual messages, utilizing trusted and respected messengers, creating and invoking social norms, identifying core goals and commitments, and providing and highlighting salient information. Specific fields of practice relevant to climate change concerns were assessed for the possible effectiveness of introducing nudging interventions: energy consumption, water use, waste and recycling, personal transportation, air travel and tourism, shopping behavior, meat consumption and dietary preferences (Bornemann & Burger, 2019; Byerly et al., 2018; Carlsson et al., 2019; Howlett & Rawat, 2019; Lindahl & Stikvoort, 2015; Mont et al., 2014; Nielsen et al., 2016; Nisa et al., 2019). Branded nudging models and mnemonics, typically found in business marketing literature, have been widely produced: SNAP (Salience, Norms, Affect, Priming); SIMPLER (Social Influence, Implementation Prompts, Mandated Deadlines, Personalization, Loss Aversion, Ease, Reminders); MINDSPACE (Messenger, Incentives,

Norms, Defaults, Salience, Priming, Affect, Commitments, Ego) (Madi, 2020, p. 38; Shankar & Foster, 2016).

Does any of this really work? This is the million dollar question. The short answer is that, at the local level and for brief periods of time, nudging and other individual behavioral change interventions clearly can have an impact. Doing things like changing dinner plate sizes, rewriting restaurant menus, resetting energy supply default options, or promoting new social norms around car use have all been shown, in highly localized studies, to have measurable effects on individuals' choices and actions that could lead to a reduction in greenhouse gas emissions. Some types of nudges—such as changes to the local physical environment or default option settings—are claimed to be more consistently effective than others (Bornemann & Burger, 2019; Mont et al., 2014). Not surprisingly, for example, "when recycling bins are located on every floor of a building, rather than only at the entrance, people are more apt to recycle because they have less distance to walk" (Nisa & Bélanger, 2019). However, whether the effects of nudges have a cumulative, enduring and positive effect at a global level, across diverse cultural and economic contexts, and over the long run, is more questionable. In a meta-analysis of eighty three randomised controlled trials on the efficacy of green nudges for addressing the climate crisis, it was found that nudges and other behavioural interventions, *when acting alone*, "promote climate change mitigation to a very small degree while the intervention lasts …, with no evidence of sustained positive effects once the intervention ends" (Nisa et al., 2019, p. 1).

It is the *when acting alone* qualification here that is perhaps the most important finding to note. For, among those who have reviewed the research with green nudges over the past decade, there is widespread agreement that nudges and other behavior change interventions can be useful complements to, but not effective substitutes for, other kinds of regulatory, infrastructural and educational investments and transformations for addressing climate crisis. "Rather than being seen as a silver bullet, the largest promise of nudge is perhaps in helping design other initiatives better and in improving the effectiveness and efficiency of policy tools and the speed of their implementation," write Mont et al. (2014, p. 8): "Nudge tools are seen as a complement to the traditional policy instruments rather than as a substitute for laws and regulations and economic tools" (see also Bornemann & Burger, 2019; Carlsson et al., 2019; Lindahl & Stikvoort, 2015; Nielsen et al., 2016; Schubert, 2017; Stern, 2020). "You can't nudge someone onto a bus service that doesn't exist," as Whitmarsh (2011) points out: "Nudge can help in [the] uptake of low-carbon alternatives, but those alternatives have to be put in place" to begin with (and we would still need to identify forms of education and intervention that are effective in promoting provision of alternatives such as accessible public transportation). The tacit learning phenomena that Thaler and Sunstein highlight with their work on nudging are important things for anyone concerned with climate change education to address, but not as the neoliberal panacea for shortcircuiting the need for strong, state-led regulatory and (formal) education interventions that Thaler and Sunstein initially suggested nudging could be.

Problems with Nudges

Problems with green (and other) nudges have been pointed out almost from the moment that Thaler and Sunstein first popularized the term, and focus on a common set of concerns: nudges are individualizing and risk shifting attention and resources away from addressing broader structural and cultural factors that are vital to tackle in effective climate change action; nudges are manipulative, infantilizing, undemocratic, uneducational or anti-educational, and quite often, unethical; nudges are simply not that effective, at least at the scale and time required for addressing the climate crisis, and at best, may help lead to local, incremental and often temporary changes in individual environmental actions and choices (Barr, 2018; Bornemann & Burger, 2019; Howlett & Rawat, 2019; Schubert, 2017; Webb, 2012). Some claims about the dangers of nudging are extreme. Furedi (2011) argues that extensive nudging will lead to the "colonisation of private life as our personal conduct becomes the target of the behaviour-management industry," and "corrosion and ultimately corruption of public life," as nudging aims to "bypass public debate and opt for psychological manipulation instead."

It is important to separate the different criticisms of green nudging. Many critics focus on the combination of the neoliberal model of promoting nudges as a substitute for regulatory, infrastructural and educational intervention to address the climate crisis, and the technocratic, surveillance capitalism model of promoting increased asymmetries of knowledge, power and learning between an elite of expert nudgers and mass of nudged humanity—what Thaler and Sunstein (2008, pp. 4–6) refer to as "libertarian paternalism." However, we need to differentiate the critique of libertarian paternalism from a critical evaluation of nudging as a theory and practice of learning and action. We certainly can and should question the promotion of green nudges as substitutes for stronger forms of cultural, political and economic action on the climate crisis. We can and should also question the undemocratic embrace of technocracy and elite asymmetries of power and knowledge. Here, alternative models can help us think through what more democratic, equitable and inclusive approaches that embrace the insights of nudging theory and practice might look like.

What, then, of the other two claims of Thaler and Sunstein's work on nudges: the importance of informal/tacit learning, and local environmental contexts in shaping this learning, as central factors that any effective climate change education and action approach must figure out how to deal with? One of the most important critiques of green nudges is that these nudges are often limited in their effectiveness because of a failure to develop a rich understanding and analysis of learning contexts. Nudges that might be demonstrated to work in controlled intervention settings often don't work, or don't work in the same way, for broader populations across time and space (Mont et al., 2014, p. 30). This might be because of differences in cultural contexts, identities and political ideologies; because of different economic situations or infrastructural constraints; or because people are continually reflecting on and learning about their surrounding environments, and may adjust their actions and choices over time (Howlett & Rawat, 2019; Schubert, 2017). Critics warn of harmful spillover or backlash effect from nudges as well: Danish executives led to eat smaller portions due to reduced plate size at a buffet banquet may eat larger meals when they return home as a consequence; diners at Gothenburg restaurants induced to choose vegetarian meals through menu design may increase their meat intake over the rest of the week; Malmö residents participating in a "No Ridiculous Car Trips Campaign" might feel "morally 'licensed' to indulge in a high-carbon holiday because they've cycled to work more in the past year" (McLoughlin et al., 2019, p. 13). The limited and context dependent effectiveness of nudges is, in fact, a reminder that fears of mass mind controlling power of elite-designed nudges may sometimes be overblown.

The underlying problem here is the "thin understanding" and "recipe book" or "piecemeal approach" that nudging takes to the ways in which environmental contexts are seen to influence tacit, informal learning (Whitehead et al., 2019; Whitmarsh, 2011). "A good rule of thumb is to assume that 'everything matters,'" Thaler and Sunstein (2008, p. 3) had written: "Small and apparently insignificant details can have major impacts on people's behavior." Scholars of informal learning would agree. "Informal learning, unlike formal and non-formal learning which are spasmodic, situated in specific places and times," writes Rogers (2014, p. 35), "is ubiquitous, universal, and continuous," and part of the everyday "practices of living." Yet, rather than develop an approach to address this core insight about the ubiquitous nature of informal learning, Thaler and Sunstein instead tried to identify simple nudging interventions that could supposedly, in and of themselves, alter the overall course of actions and choices people make: this was their neoliberal hope for an easier, cheaper, less intrusive way to address key social problems. Recognizing that we don't do our most extensive or powerful learning in controlled formal or nonformal educational settings, but rather in the course of our everyday lives, interacting with the social and physical world around us, was never going to make things easier, cheaper or simpler: rather, it should have pushed us to realize that effective climate change education and action would need to be more comprehensive and ubiquitous than ever before.

To be effective, green nudges would need to address all or at least most of the "choice architectures" relevant to climate crisis concerns that individuals encounter in the course of their daily lives: and doing this would require extensive research, regulation and infrastructural intervention (Marteau et al., 2011). Goodwin (2012, p. 90) explains the nature and size of the challenge, in relation to the use of public health nudges:

> [W]ithout regulation to control existing environments that are shaped largely by commerce, nudging people towards healthier behaviour may struggle to make much of an impression on the scale needed to improve the population's health. The problem … is that nudging people in one direction leaves them vulnerable to being nudged back again. If we consider that for every pound spent on public health marketing there will be many more spent on getting the public to eat Walkers crisps, for example, the economics of nudge starts to look flawed.

A related problem is that in seeking to identify green nudges that can transform relevant choice architectures, the nudging approach never really develops a clear theory of why current choice architectures are organized in the way that they are. To return to the green nudges listed earlier, why are plates sizes often larger rather than smaller, why do menus often favor meat over non-meat dishes, why do power companies often offer fossil fuel rather than green energy defaults, and why do local residents often do more short distance car rather than cycling trips? Is there something that links all of these, and countless other, climate change relevant choice architectures together? This is where some sociologists argue that we need a deeper analysis of "social practices" rather than "piecemeal" accounts of individual behavior (Barr, 2018; Kurz et al., 2015; Strengers & Maller, 2015). As I suggest below, this is where the critical sociology of education concept of a hidden curriculum can be helpful as well.

A second critique of nudging as a theory and practice of action is that many nudges are uneducational or anti-educational (and, as a consequence, undemocratic), and are based on an assumption that people are "hardly educable," "non-educable" or "virtually ineducable," at least in terms of being able to consciously learn to transform their own core behaviors on their own volition (Gigerenzer, 2015, pp. 365, 366, 373). "Nudging may compromise people's *ability to form (or learn) preferences*," warns Schubert (2017, p. 338; italics in original). The reason for this is that:

> nudges work by systematically relieving agents from the need to muster mental and cognitive effort: green defaults allow you to act in a pro-environmental way without even thinking about it; framing the cafeteria experience allows you to go with the vegetables with a minimum level of

> self-control; peer comparison allows you to 'save' willpower when deciding upon your energy use, and so on. (Schubert, 2017, p. 338)

"The extensive use of nudging risks the ability of a society to deliberate and learn," argue Bornemann and Burger (2019, p. 219), for "if individuals are constantly relieved of making meaningful decisions in their daily lives, they and ultimately society as a whole may lose the ability to make sound and socially robust judgments about complex societal issues." In part, this critique of nudges as uneducational or anti-educational is a critique of the neoliberal, technocratic model of nudging: but it is also a more general critique of nudging as an overall theory and practice.

There are two parts to this more general critique of nudging as an anti-educational or uneducational practice: nudges are uneducational because they eschew rational, deliberative and critically reflective approaches to teaching people about social problems like climate crisis; nudges are also uneducational because they allegedly seek to exploit cognitive weakness and human fallibility, in order to shift people's choices and actions to address these problems. "In essence, nudges are aimed at driving people's behaviours by making use of their cognitive limitations," write Ferrari et al. (2019, p. 185), "instead of enhancing their ability to make informed, rational and fully conscious decisions." "Nudge's foregrounding of the fallible, emotive human," writes Leggett (2014, p. 10), "gives up on—and possibly undermines—the rational, deliberative citizen." But there are problems with this critique. Nudges seek to work with tacit, informal forms of learning, or learning that is shaped by everyday social, physical, visual, verbal and textual environments (or choice architectures). It is true that Thaler and Sunstein emphasize the more negative aspects of this kind of learning; but equally, it needs to be remembered that this is one of the most pervasive and powerful forms of learning that exists.

The other problem with the "nudges are anti-educational" critique is that it risks embracing a narrow, traditionalist understanding of what should count as "education." For the critique looks suspiciously like a claim that promotes the superiority and preferability of formal and nonformal over informal education: after all, informal learning does not tend to occur through rational, deliberative, and critically reflective pedagogy and curriculum in the way that formal and nonformal education often do. But this promotes a romanticized view of the power of formal education to effectively address problems like climate crisis; and it fails to challenge an overly negative view of the potentialities of informal learning, which, in any event, will continue to occur ubiquitously, regardless of whether we ignore it or not. Informal learning, as Schugurensky (2006, p. 163) points out, is "often undervalued," left "at the margins of the margins of the educational conceptual and research radar," and "seldom recognized." As a consequence, this critique fails to help us address the knowledge-action gap that has long plagued formal and nonformal environmental and social justice education. Any effective climate change education program will need to embrace both formal and informal learning, while recognizing that both types of learning "have limitations and both have values: both have the potential to be emancipatory and both can be oppressive" (Rogers, 2014, p. 64). Thaler and Sunstein's work on nudging has been incredibly effective in focusing attention on the importance of informal learning for addressing the climate crisis and other pressing social problems. But to go beyond the many limitations of the nudging approach, it is helpful to turn to older traditions in education studies.

Critical Sociology of Education and the Hidden Curriculum

In the 1970s in the United States, educationalists faced a social justice conundrum. Though they had managed to reform the official curriculum in many schools to promote equal opportunity, justice and inclusion for all students, regardless of race, class or sex, the education system still seemed to reproduce clear patterns of race, class and sex inequality, injustice and exclusion. There were many reasons for this. But one key, enduring concept that education researchers and theorists invoked at the time, by way of explanation, was the idea of the hidden curriculum (Giroux & Penna, 1979). The hidden curriculum refers to "those unstated norms, values, and beliefs embedded in and transmitted to students through the underlying rules that structure the routines and social relationships in school and classroom life" (Giroux, 1983, p. 47). The argument was that, while official curricula might promote equality and justice for all, the hidden curriculum of school often worked in the opposite direction, to reproduce social inequality and injustice—and for most students, the lessons of the hidden curriculum were much more powerful than those encountered in official school lessons (Apple, 2019; Bowles & Gintis, 1976; Gillborn, 1992; Lynch, 1989). The hidden curriculum, in other words, was an attempt to offer one possible explanation for the pervasive knowledge-action gap in social justice education.

The hidden curriculum constitutes a precursor to the concept of choice architecture Thaler and Sunstein developed years later. Like choice architecture, the idea of the hidden curriculum highlights the importance of tacit, informal learning (or socialization) that occurs alongside and often in direct conflict with formal learning processes (Schugurensky, 2000); also like choice architecture, it focuses on how all aspects of local environmental context can shape students' tacit learning, from physical architecture to the organization of time schedules, social practices, school rules, student-teacher roles and relationships, and verbal, visual and textual cues (Ballantine et al., 2017; Meighan & Siraj-Blatchford, 1997). Though the original work on the hidden curriculum focused on schools, hidden curricula exist everywhere. "It is not just formal educational settings which have hidden curricula," writes Martin (1976, p. 146): "Any setting can have one and most do," as "hidden curricula can be found anywhere learning states are found." While critical sociologists were concerned with identifying hidden curricula that were unjust and problematic, since they fostered "harmful learning states," and instilled "beliefs, attitudes, values, or patterns of behavior which are undesirable," the hidden curriculum is not necessarily always a negative thing (p. 145). Some hidden curricula may foster "learning states" that are positive, or at least "neutral." "It is always possible," Martin (1976, p. 144) notes, "that we will want to embrace rather than abolish the hidden curriculum we find."

Though it has similarities to the idea of choice architecture, the hidden curriculum concept also offers several critical advantages. First, the concept assumes that there are systematic patterns that link key aspects of local (school) environments with hegemonic discourses or cultures or economic formations in society. Schools in a capitalist society, for example, may foster values and practices of "competitive individualism" in everything from their assessment models to classroom activities, school yearbooks, student council systems and extra-curricular sporting events—even as their official curriculum may call for a spirit of collective solidarity (Lynch, 1989). Where do such patterns come from? In contemporary times, there may be no deliberate intentionality or active conspiracy to import these values and practices: rather, competitive individualism has become second nature and common sense in capitalist society, an ensconced culture that many of us embrace without thinking (Hiebert, 2012). Historically, however, Vallance (1974) argues that what is a hidden curriculum today may have been part of an explicit, deliberate and intentional curriculum in earlier periods, when schools were being developed to support the growth and legitimacy of an emerging capitalist society. The "hidden" nature of this curriculum now is a marker of success of these earlier interventions in making capitalist values and practices a normative part of schooling and society. It is in this sense that Meighan and Siraj-Blatchford (1997, p. 69) describe the hidden curriculum as constituting a "hardened history" driven by "the ghosts of the architects":

> Each of the artefacts and environments that we encounter in our lives today were produced as an expression of the values of their creator/designers. In the sociology of technology this is referred to as hardened history. … The classroom may be said to be haunted by the architects who designed it and their advisers. Attempts to reshape the classroom have to fight the ideas of the originators.

A second difference from the nudging/choice architecture framework is that hidden curriculum analysis focuses more on "tacit teaching" than on "tacit learning" (Burbules, 2008; Giroux & Penna, 1979; Hiebert, 2012). If students and others tacitly learn from their local environments to embrace sexist and racist values and practices—or, in the context of the climate crisis, to act in unsustainable ways—the response of the hidden curriculum analyst is not to investigate (and assign responsibility to) individual irrationality and cognitive failure, but instead develop a critical, sociological, historical and discursive analysis of how these values and practices may be fostered (or tacitly taught) by local institutions and broader social formations. Whereas a nudging framework works in an individualizing way, turning our analytic gaze inward to look at individual choices and actions, the hidden curriculum framework moves our analytic gaze outward, to focus on social, historical and economic patterns and structures instead.

Finally, there is the question, as Martin (1976) asks, of "what should we do with a hidden curriculum when we find one?" For critical sociologists of education, the answer has been consistent: we should teach students and others to develop their critical awareness of the existence and operation of hidden curricula in all areas of their lives; and collectively, we should seek to challenge and transform hidden curricula wherever these are found to be harmful and unjust (Portelli, 1993). Giroux (1988, p. 51) argues that:

> By making both students and teachers aware of the hidden curriculum as it has traditionally operated, both groups can develop an understanding of its components and effects and work to form new insights about it. Once the hidden curriculum becomes obvious, students and teachers will be more sensitive to recognizing and altering its worst effects and can work to build new structures, methods, and social relationships.

Such "consciousness raising" as Martin (1976, p. 148) notes, does not "guarantee that a person will not succumb to a hidden curriculum"; but "one is in a better position to resist if one knows what is going on." When problematic hidden curricula are identified, the solution is not to get rid of any kind of hidden curriculum altogether: this would be impossible, as every setting has a hidden curriculum that tacitly teaches or fosters certain values, attitudes and practices among its participants. Instead, the aim is to transform these hidden curricula so that they foster desired values, attitudes and practices; *and* seek to make students and others as aware as possible of the core values and aims that have been embedded (hidden) in local contexts (Martin, 1976; Portelli, 1993). The social politics of hidden curriculum analysis are about as far as one can get from the neoliberal, technocratic, surveillance capital model of nudging: and yet, arguably, the theory and practice involved in analysing and transforming the hidden curriculum are not really different in essence from the project of creating nudges to alter choice architectures.

The hidden curriculum concept has received limited attention from those seeking to develop effective climate change education programs. A few studies note how the hidden curriculum of universities can contradict formal sustainability curricula and commitments of these institutions, as "students receive incompatible messages about sustainability from different dimensions of their experiences" at university (Winter & Cotton, 2012, p. 786). These studies suggest that "helping students deconstruct the hidden campus curriculum" on sustainability can be a powerful, transformative form of climate change learning and action (Winter & Cotton, 2012, p. 783). But to be productive, hidden curriculum analysis projects need to extend beyond formal education settings. The challenge is to engage in collective, critical reflection on how a hidden curriculum of fossil fuel based culture has come to shape virtually all aspects of daily existence, encouraging us to embrace lives of comfort, privilege and freedom fuelled by oil (e.g., Huber, 2013), as a first step toward challenging this culture by seeking to transform its local, tacit curricular manifestations. Through the concept of the hidden curriculum, there is a line we can trace from the micro level investigations of nudges embedded in local choice architectures, to the macro level analysis of the "technological, social and economic systems" (car-based transport systems, centralized electricity network systems, industrial agriculture systems, etc.) that, Pirani (2018) argues, shape and direct individual consumers' local choices around continuing fossil fuel consumption.

Progressive Education and the Third Teacher

The lineage of nudging is usually seen as coming from consumer marketing and the practice of "corporations, retailers, and modern media" trying to "manipulate … consumers through the subliminal messages of advertising that [aim] at influencing decision-making processes" (Madi, 2020, p. 79). The embrace of nudging in efforts to address social problems in education, health, welfare and the environment tends to get framed as constituting part of the commercialization and marketization of the public sphere in neoliberal society. With nudging, we find marketing experts coming into primary schools to promote marketing techniques to transform the education of young children (Charry & Parguel, 2019).

Yet, there is a markedly different lineage in education that has also long embraced nudging type insights and practices: progressive education. John Dewey wrote in 1938 of the importance of "collateral learning," referring to the importance of the "formation of enduring attitudes, of likes and dislikes" through tacit, informal learning (p. 48). Dewey (1938, p. 40) also writes what can be taken as an early call to use nudges and changes in choice architecture (or the hidden curriculum) to support progressive educational agendas:

> A primary responsibility of educators is that they not only be aware of the general principle of the shaping of actual experience by environing conditions, but that they also recognize in the concrete what surroundings are conducive to having experiences that lead to growth. Above all, they should know how to utilize the surroundings, physical and social, that exist so as to extract from them all they have to contribute to building up experiences that are worthwhile.

While the critical sociology of education is useful for developing a critical analysis of hidden curriculum (or choice architecture), it is generally less helpful for developing ideas for how to actively transform this hidden curriculum in positive ways. Progressive education, following Dewey's call to utilize physical and social surroundings in support of educational projects, can provide clear insights.

Progressive education is marked by an orientation to experiential learning, that recognises the importance of direct, practical experience as a basis for learning, and physical and social environments for shaping student experience; it is also characterized by a holistic approach that focuses on teaching the "whole person," aiming to shift not just the "head," but the "heart" and "hands" as well (Bruce & Eryaman, 2015; Little & Ellison, 2015). As a consequence, progressive educators have tended to be attentive to the need to radically transform not just the official curriculum of traditional schools and classrooms, but their hidden curriculum as well. Indeed, despite the many different forms it has taken over the years, progressive education may be defined as a critical and systematic attempt to rethink and rework the "core grammar" or hidden curriculum of traditional schooling (Darling & Nordenbo, 2003).

There are a number of traditions within progressive education that have developed models for transforming choice architecture and the hidden curriculum. Rudolf Steiner, who developed Waldorf (or Steiner) schools, began with the premise that social and physical environments inevitably "stimulate thinking" and "shape moral possibilities," and despaired at the ways in which traditional education environments often seem to undercut student learning:

> How it is to be deplored … that the schoolrooms for our children are a veritably barbaric environment for their young hearts and minds. Imagine every schoolroom … shaped by an artist in such a way that each single form is in harmony with what his [*sic*] eye should fall upon when the child is learning his [*sic*] tables. (Uhrmacher, 2004, pp. 97, 108)

In Waldorf schools, no element of the environment is "left to chance," as everything from the shape of the walls and color of paint to the organization of the timetable or extensive use of music and art in teaching core subjects is designed to be in congruence with the overall aims and vision of the formal curriculum—in effect, seeking to "nudge" the learning of Waldorf students in clearly specified directions (Hiebert, 2015, p. 33). As Kraftl (2006) notes, "ideas and ideals" of childhood and education are literally built into the physical, temporal and social organization of each Waldorf school, with the aim that this will help "evoke" among students and staff the kinds of roles, relationships, habits and practices that best embody these ideas and ideals.

Maria Montessori, who developed the worldwide network of Montessori schools, invoked the concept of the "prepared environment," in which learning aims are embedded within physical objects and the overall material organization of the school classroom (Gutek, 2004; Lillard, 2005). For Montessori, the environment exists in an interactive relationship with both the teacher and student, "with no one element at the center" (Kuh & Rivard, 2014, p. 15). The prepared environment in a Montessori school is intended to inspire and guide (or "nudge") "choice-making on the part of the child," so that children can come to learn desired lessons on their own terms, without the direct intervention of a teacher (Kuh & Rivard, 2014, p. 17). As in a Waldorf school, virtually everything in a Montessori school environment is deliberately designed to foster a form of tacit, informal learning that may be unintentional on the part of the learner, but is clearly intentional on the part of the learning provider. Furniture is small, lightweight and moveable, in order that children can learn to develop control over their environment and physical movement through space, moving tables and chairs together and apart as needed (Lillard, 2005, pp. 92–93). Floors and furniture are deliberately light-colored, so that they "would show each and every spill and smear," and "students would learn to begin to care for their environment by being active participants in the upkeep" (Jones, 2012, p. 33; see also Lillard, 2005, p. 302). Learning materials make extensive use of color-coding, so that children can "easily see what is before them and … see the patterns and categories of materials in the room" (Kuh & Rivard, 2014, p. 16).

Loris Malaguzzi, who helped to create the Reggio Emilia model of education, coined the evocative concept of the social and physical environment as being a "third teacher," in addition to actual teachers and the student her or himself (Strong-Wilson & Ellis, 2007). "It is necessary to keep in mind how influential the environment is with regard to the affective, cognitive, and linguistic acquisitions," Malaguzzi argues, noting that intended lessons can be "facilitated or obstructed by the environment and its characteristics" (quoted in Gandini, 1998, p. 166). The Reggio approach is marked by an attempt "to integrate the educational program with … the environment so as to allow for maximum movement, interdependence, and interaction" (Malaguzzi, 1998, p. 63). In Reggio schools, the learning environment is constructed as "a sort of aquarium that mirrors the ideas, values, attitudes, and cultures of the people who live within it" (Malaguzzi, quoted in Gandini, 1998, p. 177). The idea is that the school environment "reflects the vision of those who inhabit it and … shapes those visions" (Dudek, 2005, p. 50). As with Waldorf and Montessori schools, this is done through everything from the design of school architecture to class timetables to the construction of teacher-student-family-community roles and relationships. The extensive use of public displays of student art in Reggio schools seeks not just to document the student learning process, but shape it as well:

> The provision of stimulating and provocative displays of objects, whether a careful arrangement of seashells or the positioning of a mirrored surface, serves to educate children's attention to design, detail, or difference; and to contribute to the development of an alert and active response to the world. The frequent display of collective efforts, such as the woven bags of leaves collected on a walk outside, whereby each child can identify his or her contribution to the collaborative product, heightens the child's awareness of self as a member of and contributor to the larger group. (New, 1998, p. 267)

Such use of art displays in public spaces to shape desired learning outcomes is very much in line with the kinds of green nudges discussed at the outset of this chapter—changing plate size and menu design, altering printing and power supply defaults, launching a no ridiculous car trip campaign—even if the original provenance of these educative strategies is strikingly different.

The rich tradition of progressive education offers an alternative starting point for thinking about how we might embrace the most important insights of the nudge and choice architecture framework, while avoiding some of its problematic shortfalls. As with nudging, progressive education argues that educators must pay attention to tacit, informal learning, and focus on transforming social, physical and textual environments (the hidden curriculum) if they want to be effective. But progressive education differs from Thaler and Sunstein's nudge model in at least three key aspects. Progressive education does not seek to substitute forms of tacit, informal teaching and learning for the more reflexive, deliberative and rational approaches of formal education. Rather, it recognizes the problem, highlighted in the critique of the hidden curriculum, that formal and informal learning are often in conflict with one another, and seeks to ensure these forms of learning are brought into alignment, as far as possible, with a common set of explicitly defined and recognized educational goals (Martin, 1976; Rogers, 2014). Further, in embracing concepts such as the prepared environment and third teacher, progressive educators are not simply seeking to change individual student choices and actions on a piecemeal basis, as often happens with the nudging approach. Instead, these educators are guided by the principle of prefigurative utopian practice, in which "education through its processes, the experiences it offers, and the expectations it makes, should prefigure, in microcosm, the more equal, just and fulfilling society" that this education aims to (eventually) bring about for the world as a whole (Dale 1988, as quoted in Fielding & Moss, 2010, p. 148). In other words, progressive educators seek to construct learning environments in the school that reflect, embody, and call into being an overall vision and set of values, practices, habits, identities and relationships that befit an ideal future world (sustainable, carbon neutral, socially just, etc.) that they hope will be able to replace the unjust, unsustainable world that currently exists.

While some forms of progressive education have received the same concerns of elite, expert control and manipulation as Thaler and Sunstein's nudge approach (Gutek, 2004), progressive education, at its best, has been at the center of a commitment to develop a strong democratic model of learning, in which students, teachers and community members participate actively and equally in shaping educational structures, processes, aims and aspirations (Apple & Beane, 2007; Knoester, 2012; Moss, 2014). Indeed, the embrace of the environment as a third teacher in Reggio Emilia is seen as a vital part of making Reggio learning aims and values fully "transparent" and "visible" in the built environment itself, "so that they can be shared, discussed, reflected upon and, if necessary, evaluated" by "everyone—children, teachers, auxiliary staff, families, administrators and other citizens," transforming the school into a "meeting place of co-construction" and a "place of democracy" (Fielding & Moss, 2010, p. 7). It is not about seeking to hide these aims and values behind the back of Reggio learners, to slip them in surreptitiously beneath learners' conscious awareness.

At the same time, there remains something that progressive education can learn from Thaler and Sunstein's nudge approach—and indeed, from its own long history of frustration and failure in spurring effective and widespread social change. One limitation of progressive education models, whether these be from Waldorf or Montessori or Reggio Emilia or elsewhere, is that they remain focused on transforming learning environments within formal education settings. The progressive education theory of social change is to create prefigurative utopian spaces within the school, and hope that these spaces can serve as model and inspiration for broader transformations elsewhere (Fielding & Moss, 2010; Hatcher, 2007; Hoare, 1965). What the nudging approach helps to highlight is that if we want to develop an effective model of climate change (or social justice) education and action, we need to take the insights and practices of progressive education regarding prepared environments and third teachers, and apply these across all kinds of social institutions. For, once we recognise that tacit, informal learning is occurring everywhere, in all walks of life, we must acknowledge that if we wish to reshape and redirect this type of learning, we need to find ways to extend progressive theory and practice of tacit teaching beyond the walls of the progressive school. Of course, once we start trying to do this, we quickly run up against the same problem that was raised earlier regarding nudges, which is the need to tackle entrenched elite and corporate agendas and interests that often seek to nudge or teach individuals in quite different, unsustainable and climate harming directions: this is a problem of social power that neither nudging nor progressive education can help us much with at all.

References

Apple, M. (2019). *Ideology and curriculum* (4th ed.). Routledge.

Apple, M., & Beane, J. (Eds.). (2007). *Democratic schools: Lessons in powerful education.* Heinemann.

Ballantine, J., Hammack, F., & Stuber, J. (2017). *The sociology of education* (8th ed.). Routledge.

Barr, S. (2018). Environmental behaviour change. In N. Castree, M. Hulme, & J. Proctor (Eds.), *Companion to environmental studies* (pp. 748–753). Routledge.

Bobrow, E. (2018). Fight climate change with behavior change. *Behavioral Scientist*, 16 October.

Bornemann, B., & Burger, P. (2019). Nudging to sustainability? Critical reflections on nudging from a theoretically informed sustainability perspective. In

H. Straßheim & S. Beck (Eds.), *Handbook of behavioural change and public policy* (pp. 209–226). Edward Elgar.
Bowles, S., & Gintis, H. (1976). *Schooling in capitalist America.* BasicBooks.
Bruce, B., & Eryaman, M. (2015). Introduction: The progressive impulse in education. In M. Eryaman & B. Bruce (Eds.), *International handbook of progressive education* (pp. 1–51). Peter Lang.
Burbules, N. (2008). Tacit teaching. *Educational Philosophy and Theory, 40*(5), 666–677.
Byerly, H., et al. (2018). Nudging pro-environmental behavior: Evidence and opportunities. *Frontiers in Ecology and the Environment, 16*(3), 159–168.
Carlsson, F., et al. (2019). *Nudging as an environmental policy instrument.* University of Gothenburg.
Charry, K., & Parguel, B. (2019). Educating children to environmental behaviours with nudges. *Environmental Education Research, 25*(10), 1495–1509.
Darling, J., & Nordenbo, S. (2003). Progressivism. In N. Blake et al. (Eds.), *The Blackwell guide to the philosophy of education* (pp. 288–308). Blackwell.
Dewey, J. (1938). *Experience & education.* Simon & Schuster.
Dudek, M. (2005). *Children's spaces.* Architectural Press.
Feitsma, J. (2018). The behavioural state: Critical observations on technocracy and psychocracy. *Policy Sciences, 51*, 387–410.
Ferrari, L., et al. (2019). Can nudging improve the environmental impact of food supply chain? A systematic review. *Trends in Food Science & Technology, 91*, 184–192.
Fielding, M., & Moss, P. (2010). *Radical education and the common school.* Routledge.
Furedi, F. (2011). Defending moral autonomy against an army of nudgers. *Spiked*, 20 January. https://www.spiked-online.com/2011/01/20/defending-moral-autonomy-against-an-army-of-nudgers/
Gandini, L. (1998). Educational and caring spaces. In C. Edwards, L. Gandini, & G. Forman (Eds.), *The hundred languages of children* (2nd ed., pp. 161–178). Ablex.
Gigerenzer, G. (2015). On the supposed evidence for libertarian paternalism. *Review of Philosophy and Psychology, 6*, 361–383.
Gillborn, D. (1992). Citizenship, 'race' and the hidden curriculum. *International Studies in Sociology of Education, 2*(1), 57–73.
Giroux, H. (1983). *Theory and resistance in education.* Bergin & Garvey.
Giroux, H. (1988). *Teachers as intellectuals.* Bergin & Garvey.

Giroux, H., & Penna, A. (1979). Social education in the classroom: The dynamics of the hidden curriculum. *Theory & Research in Social Education, 7*(1), 21–42.

Goodwin, T. (2012). Why we should reject 'nudge'. *Politics, 32*(2), 85–92.

Gravert, C., & Kurz, V. (2021). Nudging à la carte: A field experiment on climate-friendly food choice. *Behavioural Public Policy, 5*(3), 378–395.

Gutek, G. (2004). Introduction: A biography of Montessori and an analysis of the Montessori method. In G. Gutek (Ed.), *The Montessori method* (pp. 1–66). Rowman & Littlefield.

Hansen, P. (2013). Nudge-experiment: How to reduce food waste among CEOs. *The iNudgeyou blog*. https://inudgeyou.com/en/nudge-experiment-how-to-reduce-food-waste-among-ceos/

Hatcher, R. (2007). Yes, but how do we get there? Alternative visions and the problem of strategy. *Journal of Critical Education Studies, 5*(4), 78–109.

Hiebert, M. (2012). Pulling back the curtains of the hidden curriculum. *Aroundsquare*, 4 May. https://www.aroundsquare.com/around/2012/5/4/pulling-back-the-curtains-of-the-hidden-curriculum.html

Hiebert, M. (2015). *Tacit teaching and the educative context: Bringing intentionality into the hidden curriculum*. PhD Dissertation, University of Calgary.

Hoare, Q. (1965). Education: Programmes and men. *New Left Review, 32*, 40–52.

Howlett, M., & Rawat, S. (2019). Behavioral science and climate policy. In *Oxford research encyclopedia, climate science*. https://doi.org/10.1093/acrefore/9780190228620.013.62410

Huber, M. (2013). *Lifeblood: Oil, freedom and the forces of capital*. University of Minnesota Press.

IPCC—Intergovernmental Panel on Climate Change. (2018). *Global warming of 1.5°C*. IPCC.

Jones, S. (2012). *Places of purpose, purposeful places: An exploration of the physical environment of primary level classrooms*. PhD Dissertation, University of Denver.

Jones, R., Pykett, J., & Whitehead, M. (2013). *Changing behaviours: On the rise of the psychological state*. Edward Elgar.

Knoester, M. (2012). *Democratic education in practice*. Teachers College Press.

Kollmuss, A., & Agyeman, J. (2002). Mind the gap: Why do people act environmentally and what are the barriers to pro-environmental behavior? *Environmental Education Research, 8*(2), 239–260.

Kraftl, P. (2006). Building an idea: The material construction of an ideal childhood. *Transactions of the Institute of British Geographers, 31*, 488–504.

Kuh, L., & Rivard, M. (2014). The prepared environment: Aesthetic legacies of Dewey, Montessori, and Reggio Emilia. In L. Kuh (Ed.), *Thinking critically about environments for young children* (pp. 11–32). Teachers College Press.
Kurz, T., et al. (2015). Habitual behaviors or patterns of practice? Explaining and changing repetitive climate-relevant actions. *WIREs Climate Change, 6*, 113–128.
Leggett, W. (2014). The politics of behaviour change: Nudge, neoliberalism and the state. *Policy & Politics, 42*(1), 3–19.
Lenhart, J. (2013). Malmö: No ridiculous car journeys. *The Urban Observer*, 25 May. https://exploring-and-observing-cities.org/2013/05/25/malmo-no-ridiculous-car-journeys/
Lillard, A. (2005). *Montessori: The science behind the genius.* Oxford University Press.
Lindahl, T., & Stikvoort, B. (2015). *Nudging—The new black in environmental policy?* FORES.
Little, T., & Ellison, K. (2015). *Loving learning: How progressive education can save America's schools*. W.W. Norton.
Lynch, K. (1989). *The hidden curriculum*. Falmer Press.
Madi, M. (2020). *The dark side of nudges*. Routledge.
Malaguzzi, L. (1998). History, ideas, and basic philosophy: An interview with Lella Gandini. In C. Edwards, L. Gandini, & G. Forman (Eds.), *The hundred languages of children* (2nd ed., pp. 49–98). Ablex.
Maniates, M. (2001). Individualization: Plant a tree, buy a bike, save the world? *Global Environmental Politics, 1*(3), 31–52.
Marteau, T., et al. (2011). Judging nudging: Can nudging improve population health? *British Medical Journal, 342*, 263–265.
Martin, J. (1976). What should we do with a hidden curriculum when we find one? *Curriculum Inquiry, 6*(2), 135–151.
McLoughlin, N., et al. (2019). *Mainstreaming low-carbon lifestyles.* Climate Outreach.
Meighan, R., & Siraj-Blatchford, I. (1997). *A sociology of educating* (3rd ed.). Cassell.
Mont, O., Lehner, M., & Heiskanen, E. (2014). *Nudging: A tool for sustainable behaviour?* Swedish Environmental Protection Agency.
Moss, P. (2014). *Transformative change and real utopias in early childhood education*. Routledge.
New, R. (1998). Theory and praxis in Reggio Emilia: They know what they are doing, and why. In C. Edwards, L. Gandini, & G. Forman (Eds.), *The hundred languages of children* (2nd ed., pp. 261–284). Ablex.

Nielsen, A., et al. (2016). *Nudging and pro-environmental behaviour*. Nordic Council of Ministers.
Nisa, C., & Bélanger, J. (2019). Can you change for climate change? *Scientific American*, 23 December. https://blogs.scientificamerican.com/observations/can-you-change-for-climate-change/
Nisa, C., et al. (2019). Meta-analysis of randomised controlled trials testing behavioural interventions to promote household action on climate change. *Nature Communications, 10*(4545), 1–13.
OECD. (2017a). *Behavioural insights and public policy: Lessons from around the world*. OECD.
OECD. (2017b). *Tackling environmental problems with the help of behavioural insights*. OECD.
Pichert, D., & Katsikopoulos, K. (2008). Green defaults: Information presentation and pro-environmental behaviour. *Journal of Environmental Psychology, 28*, 63–73.
Pirani, S. (2018). *Burning up: A global history of fossil fuel consumption*. Pluto.
Portelli, J. (1993). Exposing the hidden curriculum. *Journal of Curriculum Studies, 25*(4), 343–358.
Rogers, A. (2014). *The base of the iceberg: Informal learning and its impact on formal and non-formal learning*. Barbara Budrich.
Rutter, T. (2016). How the World Bank is 'nudging' attitudes to health and hygiene. *Guardian*, 4 March.
Schubert, C. (2017). Green nudges: Do they work? Are they ethical? *Ecological Economics, 132*, 329–342.
Schugurensky, D. (2000). *The forms of informal learning: Towards a conceptualization of the field*. WALL Working Paper No. 19. Toronto: OISE.
Schugurensky, D. (2006). 'This is our school of citizenship': Informal learning in local democracy. *Counterpoints, 249*, 163–182.
Selinger, E., & Whyte, K. (2012). Nudging cannot solve complex policy problems. *European Journal of Risk Regulation, 3*(1), 26–31.
Shankar, M., & Foster, L. (2016). *Behavioural insights at the United Nations*. United Nations.
Solution Search. (2018). Climate change needs behavior change. https://climatecommunication.yale.edu/news-events/climate-change-needs-behavior-change/.
Solution Search. (2020). Climate change needs behavior change. https://solutionsearch.org/contests/climate-change-needs-behavior-change.
Stern, P. (2020). A reexamination of how behavioral interventions can promote household action to limit climate change. *Nature Communications, 11*, 918.

Strengers, Y., & Maller, C. (Eds.). (2015). *Social practices, intervention and sustainability: Beyond behaviour change*. Routledge.
Strong-Wilson, T., & Ellis, J. (2007). Children and place: Reggio Emilia's environment as third teacher. *Theory into Practice, 46*(1), 40–47.
Sunstein, C., & Reisch, L. (2014). Automatically green: Behavioral economics and environmental protection. *Harvard Environmental Law Review, 38*, 127–158.
Thaler, R., & Sunstein, C. (2008). *Nudge: Improving decisions about health, wealth, and happiness*. Yale University Press.
Uhrmacher, B. (2004). An environment for developing souls: The ideas of Rudolf Steiner. *Counterpoints, 263*, 97–120.
UN—United Nations. (2017). Nudge to action: Behavioural science for sustainability. 3 March. https://www.unenvironment.org/news-and-stories/story/nudge-action-behavioural-science-sustainability
UNEP—United Nations Environmenal Program. (2017). *Consuming differently, consuming sustainably: Behavioural insights for policymaking*. UNEP.
Vallance, E. (1974). Hiding the hidden curriculum: An interpretation of the language of justification in nineteenth-century educational reform. *Curriculum Theory Network, 4*(1), 5–21.
Webb, J. (2012). Climate change and society: The chimera of behaviour change technologies. *Sociology, 46*(1), 109–125.
Whitehead, M., et al. (2019). Neuroliberalism: Cognition, context, and the geographical bounding of rationality. *Progress in Human Geography, 43*(4), 632–649.
Whitmarsh, L. (2011). The problems with nudge. *Green Alliance Blog*, 31 March. https://greenallianceblog.org.uk/2011/03/31/the-problems-with-nudge/
Williamson, K., Satre-Meloy, A., Velasco, K., & Green, K. (2018). *Climate change needs behavior change: Making the case for behavioral solutions to reduce global warming*. Rare.
Winter, J., & Cotton, D. (2012). Making the hidden curriculum visible: Sustainability literacy in higher education. *Environmental Education Research, 18*(6), 783–796.
World Bank. (2015). *World development report 2015: Mind, society, and behavior*. World Bank.
Zuboff, S. (2019). *The age of surveillance capitalism*. PublicAffairs.

4

Pedagogies of Hope and Fear

In January 2019, Greta Thunberg, a sixteen year old Swedish climate activist who had become internationally known for helping to launch the Fridays for Future school strike for climate movement, went to the annual gathering of the world's business and political elite in Davos, Switzerland, to deliver a speech on the climate crisis. At the end of her speech, Thunberg (2019a) complained that "adults keep saying: 'We owe it to the young people to give them hope.'" "But I don't want your hope," Thunberg retorted, "I don't want you to be hopeful. I want you to panic. I want you to feel the fear I feel every day." It is a theme that Thunberg has come back to often. In April 2019, Thunberg went to Strasbourg to speak in the EU Parliament and began with these words: "My name is Greta Thunberg. I am 16 years old, I come from Sweden, and I want you to panic. I want you to act as if your house was on fire" (Thunberg, 2019b). In January 2020, Thunberg returned to Davos to repeat her message:

> One year ago, I came to Davos and told you that our house is on fire. I said I wanted you to panic. … You say children shouldn't worry. You say, 'Just leave this to us. We will fix this. We promise we won't let you down. Don't

S. Tannock, *Educating for Radical Social Transformation in the Climate Crisis*, Palgrave Studies in Education and the Environment,
https://doi.org/10.1007/978-3-030-83000-7_4

> be so pessimistic. … [But] our house is still on fire. Your inaction is fuelling the flames by the hour. (Thunberg, 2020)

In April 2020, as the world was battling a coronavirus pandemic, Thunberg's climate action organization, Fridays for Future (2020), released a video for the fiftieth annual Earth Day, that sought to instill a sense of panic and fear among its viewers. In the video, titled *Our House is on Fire*, a family gets up in the morning, has a cheerful breakfast together, talks about an upcoming school test for one of the children, and merrily sees the children off to school, oblivious to the fact that their home is literally on fire, with flames and smoke all around them. "Our house is on fire," the video concludes silently: "React."

Thunberg's rhetoric has been strongly contested by rightwing commentators. "We must reject the perennial prophets of doom and their predictions of the apocalypse," US President Donald Trump declared at Davos in early 2020: "This is not a time for pessimism. This is a time for optimism. Fear and doubt is not a good thought process because this is a time for tremendous hope and joy and optimism" (Euronews, 2020). Greta Thunberg is "an avatar of 'catastrophic thinking'" who "wrongly pushes a message of doom and gloom," writes Soave (2019); she is a "prophet of doom" who has "taken it upon herself to create a s***storm and get the rest of us in an apocalyptic frame of mind," complains Scicluna (2019). Such criticisms are not just limited to the climate denialist right. Among climate change educators and activists, too, there is a continuing debate about the relative importance of what Russell and Oakley (2016, p. 13) refer to as the different "emotional dimensions" of environmental education and activism: "Fear. Guilt. Shame. Anger. Disgust. Frustration. Despair. Apathy. Desire. Compassion. Empathy. Wonder. Joy. Love. Hope." An "educational movement that leaves its participants in despair, hopeless, immobilized by dread, or, so disenfranchised they resort to hyper-consumerism," warn Kelsey and Armstrong (2012, p. 190), "is neither morally defensible nor likely to lead to sustainability outcomes." Concern about the use of "fear appeals" in climate change education and communication tends to be particularly pronounced when these are targeted at young children. "No tragedies before fourth grade," insists environmental educator David Sobel, who warns that "prematurely recruiting children to solve [climate change] will just make them feel helpless and hopeless" (Sobel, 1996, p. 5; 2007, p. 15).

In this chapter, I consider the significance of hope, fear and other emotions for developing effective climate change education and action. This includes examining the key problems with invocations of hope that have raised Thunberg's ire; but also the worries about the possible harmful effects that fear, panic and anxiety in relation to the climate crisis are sometimes claimed to have. I suggest that one way out of these conflicts is to return to the theory and practice of a "pedagogy of hope," that is most closely associated with Paulo Freire, but has a long history as a core component of radical education projects committed to fundamental social change among leftist, feminist and anti-racist activists and educators. In working through these conflicts and debates over the appropriate place of hope and fear in climate change education, we need to recognize that, for climate change educators and activists, these debates are not about whether or not there is a climate crisis that we should be very worried about, nor even about emotions themselves—although, the lived experience of emotions matters very much to individuals, communities and society as a whole. Rather, underlying these debates is a concern about climate change action and social transformation, and a dispute about the links between emotion and transformative action and real social change. "I want you to panic. I want you to feel the fear I feel every day. *And then I want you to act*," is how Thunberg finished her 2019 Davos speech (emphasis added). Discussions regarding pedagogies of hope and fear are not principally about how or whether we should learn to be more hopeful or fearful in our daily lives: they are, in the end, about what and how we need to learn and be, in order to undertake effective, transformative climate change action in the world.

What's the Matter with Hope?

For the past half century, community organizations working in the tradition of Saul Alinsky have often trained new community organizers using an activity based on a chapter from Thucydides' *History of the Pelopponnesian Wars*, set in Ancient Greece, that is known as the "Melian

Dialogue." In the Dialogue, the Athenian army has just landed on the small Greek island of Melos, demanding to take control of the island as part of the Athenian Empire; a small group of Melian leaders are meeting with the Athenians to plead for their island's continued independence. The Melians make appeals to principles of justice, faith and hope, while failing to recognize the facts on the ground of being vastly overpowered by the military, political and economic might of the Athenian Empire. The Athenians, as realists and pragmatists, are scornful of the Melians' vain invocations of hope:

> Hope, that comforter in danger! If one already has solid advantages to fall back upon, one can indulge in hope. … But hope is by nature an expensive commodity, and those who are risking their all on one cast find out what it means only when they are already ruined; it never fails them in the period when such a knowledge would enable them to take precautions. … [D]o not be like those people who … miss the chance of saving themselves in a human and practical way, and, when every clear and distinct hope has left them in their adversity, turn to what is blind and vague, to prophecies and oracles and such things which by encouraging hope lead men to ruin. (Thucydides, 1974, pp. 225–226)

When the Melians refuse to listen to or speak the Athenians' language of power, the Athenians launch a military assault on the island, kill all Melian men of military age, sell the women and children into slavery, and take over Melos as a colony in its empire. Alinskyite community organizations use the discussion of the Melian Dialogue to try to shock new would-be organizers, who are often likely to identify initially with the Melian commitment to idealism, faith and justice. The intended lesson of the training is that for community organizers to be effective in taking transformative action to change society, they have to think and act less like a Melian, and discover their inner Athenians (Boyte, 2004, pp. 39–41; Bretherton, 2015, pp. 137–138; Walls, 2015, pp. 62–63). They need to start by recognizing "the world as it is," develop grounded political analyses, and build up their communities' power in order to be able to actually make real changes in society. The point is not to stop being hopeful of change in the world: but rather, to couch any such hope in sober analysis and pragmatic, strategic action. In this, the Alinsky training session on the Melian Dialogue is very much in keeping with Greta Thunberg's critique of hope invocations in the context of the contemporary climate crisis.

In the world as it exists today, there is social pressure to think like a Melian: what Lesley Head (2016, p. 2) calls "a relentless cultural disposition to focus disproportionately on positive outcomes" and not be "a doom and gloom merchant." If the climate crisis emerged during the rise of the behavioral change state, as suggested in the previous chapter, it also appeared in a period that has seen the worldwide rise to prominence of positive thinking, positive psychology and happiness studies, embraced and promoted by elites as pivotal for addressing all kinds of individual and social problems. The central claim of positive thinking, Ehrenreich (2009, pp. 334–335) observes, "is that happiness—or optimism, positive emotions, positive affect, or positive *something*—is not only desirable in and of itself but actually useful, leading to better health and greater success." Individuals are pushed to "maintain a positive attitude at all times and at all costs" (Held, 2002, p. 967). Positive thinking is tied to the hegemony of neoliberal individualism: positivity, happiness, optimism and resilience are seen as "optimal" internal "emotional states" that individuals can, with expert training, be expected to manipulate (Binkley, 2014, p. 2); social "progress comes not from changing institutions but from transforming how people see the world" (Horowitz, 2018, p. 8). A new "hope theory" has been developed to define, measure and activate "hopeful thinking," which is said "to drive the emotions and well-being of people" (Snyder et al., 2005, p. 257). Positive thinking has been described as a "cultural movement" and "cultural zeitgeist," an "industry" and "global institutional network," that has spread rapidly across the fields of "economics, education, therapeutics, health, politics, criminology, sports science, animal welfare, design, neurosciences, the humanities, management and business" (Cabanas & Illouz, 2019; Held, 2002; Horowitz, 2018). Critics point out the questionable scientific foundations of the movement, noting how transparently it serves the ideological interests of the political and economic elites who so enthusiastically support and promote its work, thanks to its individualizing rhetoric and insistence that people who are unsatisfied with the status quo should look inward at their self practices and identities, rather than contest dominant social, political and economic structures (Binkley, 2014; Cabanas & Illouz, 2019; Davies, 2015; Ehrenreich, 2009; Horowitz, 2018).

Inevitably, this elite driven emphasis on the importance of fostering positive thinking and happiness has made its way to environmental education and action on climate crisis as well. There are widespread claims that a key problem in climate change education and communication is its "doom and gloom" approach, and calls to adopt a more positive and hopeful narrative (Kelsey, 2016; Saylan & Blumstein, 2011). Environmental educators speak of prevailing expectations to "focus on the optimistic and positive," and "teach hope" (Selby & Kagawa, 2018, p. 314; Maniates, 2016, p. 137). The author of an IPCC report on the climate crisis suggests the public should "look on [the] bright side of climate change," by "staying positive about the future" and focusing on the "'really exciting opportunities' to adapt to the impacts of climate change" (Gosden, 2014). Scientists at the Smithsonian Institution in the US and Cambridge Conservation Initiative in the UK recently created the Earth Optimism Alliance, as "a worldwide movement aimed at fundamentally shifting how we frame the environmental narrative, from one of doom and gloom to one focused on solutions and successes" (Earth Optimism, 2020a; see also Brown, 2017). Earth Optimism events are now held around the world—in Cambridge, Washington, Sydney, Nairobi, Beijing and Rio de Janeiro—to "celebrate environmental success stories," "deliver stories of hope," and "focus on the positive" (Cambridge Network, 2020). "In these difficult times," a "Special Message from the Alliance" tells us, environmentalists need "optimism more than ever to encourage and guide us in our work" (Earth Optimism, 2020b).

In 2016, Christina Figueres, the former director of the United Nations Framework Convention on Climate Change, co-founded Global Optimism as a "purpose driven enterprise" that seeks "to precipitate a transformation from pessimism to optimism as a method of creating social and environmental change" (Devex, 2020). Figueres argues that addressing the problem of climate crisis depends on our ability to foster a new "mindset" and "shift in consciousness" that she calls "Stubborn Optimism": this optimism has "transformative power" and is "the force that enables you to create a new reality" (Figueres & Rivett-Carnac, 2020, pp. 68, 69, 77). Supporters of this vision can be seen wearing branded "Stubborn Climate Optimist" t-shirts; while celebrities offer soundbites to tell us that "there is no limit to what you can achieve when you change your assumption of what is possible" (Global Optimism, 2020; Mission2020, 2020). In 2020, Yale Climate Connections assembled a list of books published recently on hope and the climate crisis (Svoboda, 2020). The list, as the site noted, was "surprisingly long" and included titles such as *Climate of Hope* (Bloomberg & Pope, 2017), *Atmosphere of Hope* (Flannery, 2015), *The Hard Work of Hope* (Sandford & O'Riordan, 2017), *Where's the Hope?* (Bilodeau, 2018), *The Archipelago of Hope* (Raygorodetsky, 2017), *Hope in the Age of Climate Change* (Doran, 2017), and *The Spirit of Hope: Theology for a World in Peril* (Moltman, 2019). Reading these works, one finds most say little about the actual substance of hope in any great detail, and the approaches taken to climate change vary considerably from one book to another: "hope" seems to serve as an invocation or badge, to signal to readers that the approach presented in *this* particular book is the one most likely to offer the way forward on solving the climate crisis.

When hope becomes foundational and essential for environmental (and other forms of social justice) education and action, it becomes important to find ways to locate and measure it: who has lots of it and who has less, and how can we know if our actions are increasing or decreasing the amount of hope that individuals have. As Boyce (2014, p. 2925) observes, while hope has existed as a concept throughout human history, it is only in the past quarter century that there has been a proliferation of scales, indices and other quantitative measurements designed to measure hope precisely: the Nowotny Hope Scale, Snyder Hope Scale, Herth Hope Index, Miller Hope Scale, Zimmerman Hope Scale, Beck Hopelessness Scale, and the Integrative Hope Scale (see also Li & Monroe, 2018). There is even now a "Climate Change Hope Scale" that assesses how hopeful each of us are about global climate change, and asks respondents to agree or disagree with statements such as "I know that there are things that I can do to help solve problems caused by climate change," "I am energetically pursuing ways to solve problems caused by climate change," "If everyone works together, we can solve problems caused by climate change," or "I believe that scientists will be able to find ways to solve problems caused by climate change" (Li & Monroe, 2018; Stevenson & Peterson, 2015).

Within the field of climate change education, perhaps nobody has done more to focus critical attention on the nature and significance of hope than Maria Ojala, an environmental psychologist in Sweden. Defining hope as "an emotion that is activated by the belief and expectation that things will turn out well in the end," Ojala argues, in a series of empirical studies and conceptual discussions, that hope, in and of itself, provides no guarantee whatsoever for effective climate crisis action (Ojala, 2012b, p. 541). Indeed, just as Greta Thunberg and Alinskyite community organizers warn, hope can be a complete and total disaster. Hope in relation to the climate crisis may be based in the "denial" of reality, may be an "illusion," a "feel-good emotion" and a form of "unrealistic optimism," linked to "disengagement" and "inactivity," that becomes a way of "escaping responsibility to act to try to change the situation," and helps to preserve "the status quo" (Ojala, 2012a, p. 627; 2015, p. 133; 2017, pp. 77, 79). Some learners, Ojala (2017) observes:

> evoke hope by de-emphasizing, or even denying, the seriousness of the problem [climate change]. They claim for instance that they feel hopeful because the climate problem is exaggerated in the media, or that the problem does not concern them personally since the negative effects will only be visible in a faraway future. (p. 79)

Ojala is far from alone in making such claims. "One often encounters claims that we are living in 'unhopeful times' … and that we have lost our 'vocabulary of hope,'" Webb (2013, p. 411) writes, and "the need to develop a pedagogy of hope is presented as one of the most urgent tasks facing educators." However, Webb warns, "there is nothing inherently radical or subversive about hope or a hope-driven pedagogy," as "pedagogies of hope can operate to conserve and reproduce existing social relations as well as to transform them" (pp. 399, 412). Selby and Kagawa (2018, p. 314) likewise warn that hope in climate change education can all too often be "cosy" and "pollyannaish," a "spurious optimism, a wishful thinking, a comfortable fiction based on what we would prefer to see happen while keeping our 'eyes wide shut.'"

In her research on hope and climate change education, Ojala argues that a distinction needs to be made between good and bad forms of hope, which she labels as "constructive hope" versus "hope based on denial." Constructive hope is based on a positive reframing of the climate crisis, "trust in different societal actors" (for example, science, technology, or the government), and "trust in one's own ability to influence environmental problems in a positive direction," and is often linked positively to "pro-environmental behavior," such as recycling, buying environmentally friendly products, conserving water or energy use in the home, or cycling, walking or using public transportation instead of driving (Ojala, 2012a, pp. 628, 631). Constructive hope, Ojala (2016, p. 50) argues, should be actively encouraged by educators, because it "can give people the strength to confront and do something about" the climate crisis. But there is a problem with Ojala's constructive hope as well. For constructive hope is linked to precisely the kinds of private, individualized, consumerist environmental practices that have been widely critiqued as being, at best, limited in their ability to substantively impact the climate crisis, and at worst, a depoliticizing agenda that works to distract energy and attention from the need to collectively pressure political and economic elites for institutional and systemic change (Jorgenson et al., 2019; Maniates, 2001, 2016; Waldron et al., 2019). Trust in technology or the state to solve the problem of climate crisis can also be exactly what we don't need in effective climate change education and action: the *failure* of states to act has long been at the heart of the problem of climate crisis, and the recurrent turn to forms of technological utopianism is often a type of "magical thinking" that is poorly supported by scientific evidence, and can lead to the embrace of geoengineering proposals that risk being harmful or dangerous (Klein, 2014; Mann & Toles, 2016).

There is a danger, in other words, that climate change education that is oriented to fostering "constructive hope" among students may be effective in developing students who are hopeful about the future, enthusiastically embrace environmentalist identities, and carry out local resource and energy conservation projects—but whose actions don't actually do anything that will substantively address the climate crisis. "Saving the planet," Maniates (2016, p. 142) suggests, becomes little more than a "lifestyle choice." As Ideland (2015) argues, climate change education produces happy, empowered, "action-competent" children whose actions "are supposed to be 'up-lifting'" and "contribute to the displacement of the negative feelings that arise when subjected to looming environmental catastrophes"; yet, at the same time, these "actions often are symbolic rather than impacting the environment." We regularly see such activities celebrated in the media. In February 2020, a group of primary school children in Cork, Ireland, made international headlines after a climate change rap video they did went viral, watched over a hundred and thirty thousand times on YouTube: the video went on to win the European Schoolvision song contest, and the children were invited to appear on the Good Morning America TV show and meet Ireland's head of government, Leo Varadkar (Hilliard, 2020). The rap, titled "One Small Change," tells listeners that "we can make a difference in our own small way," by doing things like turning off the tap, cycling to school, buying local produce, and recycling plastic (Irish Times, 2020). But if small actions and creative projects like these are not linked to larger, collective, political actions to push for institutional transformation, they can be highly inadequate and even problematic.

Take the case of recycling plastic invoked by the Irish "One Small Change" primary school rappers, which Ojala (2008) also discusses in relation to climate change hope. Recycling, as Maniates (2001, p. 37) argues, is "a prime example of the individualization of responsibility." Despite widespread popularity, recycling only manages to affect a tiny proportion of the plastic produced each year, and rather than solving environmental problems, tends to export them to other countries, inflicting poor working conditions and public health hazards for those working in or living near recycling industry plants; what is more, recycling has been deliberately promoted by major corporate plastic producers as a way to avoid regulations that would restrict their ability to continue producing plastic, while shifting responsibility for the problems caused by plastic production onto individual consumers (Lerner, 2019a, 2019b; MacBride, 2012). As Samantha MacBride (2012, p. 6) argues in *Recycling Reconsidered*, recycling has been most successful in producing a condition of "busy-ness," defined as "a fulfilling sense of work and achievement that often brings positive side effects but fails to reach the central effect." Busy-ness, MacBride notes, "is a handy method of maintaining the status quo yet is simultaneously active, optimistic, and often makes people feel better." MacBride's concept of busy-ness should raise questions around common claims, in the field of climate change education, that "the benefit of action is that it is a pedagogic means for … overcoming feelings of distress, hopelessness and fear" (Stevenson et al., 2017, p. 68). A growing number of environmental organizations are now "demanding that we push past the idea of recycling and require corporations to limit plastics production" in the first place (Lerner, 2019b).

What we need is a different basis for thinking about hope, one less centered on reshaping individual emotion and promoting positivity, trust and local pro-environmental behaviors, and focused more on the need for fundamental institutional and systemic transformation. Ojala, through her continuing interrogation of the role of hope in climate change education, has herself increasingly come to recognize this need in her more recent work. The "concept of hope" has been "included in research about factors that promote or hinder private-sphere pro-environmental behaviour," writes Ojala (2017, p. 77), but it needs to be considered in a more critical and foundational sense for the role it might play in relation to "transformative learning that prepares students for societal change." To make this shift, it is useful to turn to older traditions in radical education and philosophy on critical pedagogies of hope and (real) utopian practice.

Fear and Anxiety in Climate Change Education

Greta Thunberg's recent calls for panic and fear certainly run counter to the current cultural and educational zeitgeist in the West, where aversion to "fear appeals" in climate change communication and education is strong, and forms the flipside of calls to embrace positivity and hope (Pihkala, 2017; Reser & Bradley, 2017). As González-Gaudiano and Meira-Cartea (2019, p. 397) note, "fear and despair are often considered anti-pedagogical emotions." In the 1990s, David Sobel (1996) wrote a book called *Beyond Ecophobia*, warning that children were becoming increasingly scared of environmental harm and destruction (what Sobel calls "ecophobia"), and this phobia might have a harmful impact on children's well-being and willingness and ability to take action to address the climate crisis and other environmental problems, making them feel helpless and hopeless. Since then, similar terms have appeared, as environmental educators, psychologists and researchers have raised concerns about the spread of "climate anxiety," "eco-anxiety," "climate grief," "climate distress," "climate depression," and "atmosfear," not just among children but adults as well (Bryan, 2020; Kundzewicz et al., 2020). The American Psychological Association defines eco-anxiety as "a chronic fear of environmental doom," frequently accompanied by feelings of loss, helplessness, frustration, fatalism, depression and grief (Whitmore-Williams et al., 2017, pp. 27, 68).

Studies now commonly find widespread pessimism, depression and anxiety about the climate crisis, among children and adults. In Australia, a survey of 10–14 year olds found that over four in ten children are "worried about the future impact of climate change" and a quarter fear the world will end during their own lifetimes due to the climate crisis and other threats (Tucci et al., 2007, p. 13). Another survey in Australia of high school and university students found that eight in ten students are "somewhat or very anxious about climate change, with close to half of those experiencing these emotions on a weekly basis" (Ward, 2019). Similar findings have been reported elsewhere around the world. In the UK, 17% of children are losing sleep and having their eating habits affected by anxiety over the climate crisis (Young, 2020). In the United States, 69% of adults are somewhat worried and 29% are very worried about the climate crisis, and over half (51%) feel helpless in relation to the crisis (Leiserowitz et al., 2018). In Finland, a survey found that a quarter of all Finns report suffering from feelings of climate anxiety, while smaller numbers experience numbness and depression in relation to the climate crisis (Autere, 2019). It has been suggested that one of the strongest forms of climate denial now is based not on the claim that the climate crisis does not exist, but rather on the feeling that it is too overwhelming to think or do anything about. Haltinner and Sarathchandra (2018, p. 2) call this "the 'ostrich effect,' as people metaphorically 'put their heads in the sand' to avoid uncomfortable information" (see also Norgaard, 2011). The conclusion for many is that environmental educators and communicators need to drop their habitual fear-mongering ways—warning of imminent catastrophes facing penguins, polar bears, coral reefs, rainforests, even human survival, and so on—and find ways to embrace and promote climate hope, climate positivity and climate optimism.

As with hope, these claims and concerns around fear in relation to climate change education need to be carefully unpacked. Nobody wants to see any learner, young or old, terrorized by what they learn; and if climate change education induced fear, anxiety and panic leads only to paralysis, passivity or denial among students, then clearly this is not an effective educational approach. But questions need to be asked. One is whether fear, anxiety and panic are caused by the fact and reality of the climate crisis, or primarily by the ways in which educators and communicators opt to teach and talk about this crisis. When concerns are raised about educators "presenting negative messages of 'gloom and doom,'" "conveying a negative outlook on the state of the environment," using "crisis discourse," making "fear appeals," or taking a "catastrophe approach," the suggestion is that the principal source of the problem lies in the rhetorical choices made by these educators (Hammonds, 2008; Russell & Oakley, 2016, p. 14; Selby & Kagawa, 2018, p. 314). But the climate crisis is, in and of itself, terrifying and panic-inducing (Wallace-Wells, 2019). If it exists, there is little point pretending it doesn't exist or promoting false hopes: as Alinskyite community organizers might say, this is the familiar, problematic Melian response (and look what happened to the Melians). Climate anxiety, as many note, "can be seen as a reasonable, rational response to the significant changes the world is going through" (Sutton & Smith, 2020). "In these extreme circumstances," argues Hickman (2019), "a bit of depression about the environment" is "the only sane response." Indeed, we would probably need to be more concerned if such climate anxiety and fear did *not* exist—for this would constitute part of what Amitav Ghosh (2016, p. 8) calls the "great derangement," an "imaginative and cultural failure" that is "blind to potentially life-changing threats." At the end of the day, the solution to climate change anxiety lies not in the artifice of climate change communications, but in solving the problem of the climate crisis itself.

A second question is whether climate anxiety, fear and panic lead only or always to paralysis, denial, passivity and inactivity. There is no doubt that such links can occur: and climate based fear can be extremely debilitating for some individuals. Stories have been reported of couples declining to have children due to their climate crisis fears, or a boy in Australia who refused to drink water in response to unprecedented drought (Blum, 2020; Plautz, 2020). Groups like the Climate Psychology Alliance in the UK and the Good Grief Network in the US have been created, in part, to address such problems. We can point to studies such as Kari Norgaard's (2011) *Living in Denial*, an ethnography of a community in Norway, that powerfully depicts the ways in which collective feelings of fear, guilt and hopelessness lead to a shared response of distancing, denial and inaction in relation to the climate crisis.

However, as with hope, there is no automatic or inevitable link between fear or anxiety and climate crisis paralysis and inaction. As Chapman et al. (2017, pp. 850, 852) argue, emotions such as hope or fear are not "simple levers communicators can pull to obtain specific goals," and empirical research and theory on how emotions work in society "do not support definitive, simplistic, and overly broad assertions about the effect of specific emotions on climate change responses" (see also Kidd et al., 2019; Reser & Bradley, 2017). In many cases, climate anxiety (or fear or panic etc.) can be a vital spur or resource that can inspire effective climate crisis action (Pihkala, 2019; Verplanken & Roy, 2013; Yang & Kahlor, 2012). In the UK, the environmental charity Global Action Plan released a guidebook called *Climate Anxiety: An Introduction for Teachers*: climate anxiety, the book argues, "is a real issue affecting many adults and young people today but it also has the potential to empower and motivate people towards positive action to help protect our planet" (Global Action Plan, 2020, p. 3). Ojala's (2007, pp. 741, 743) study of young environmental activists in Sweden found these activists "generally perceived their worry as an unproblematic and necessary emotion," and "both negative and positive emotions play important parts in these young people's engagement." Ojala concludes that if we want to develop effective climate change education and action, "it is not the ability to get rid of worry that should be sought after but rather the capacity to face worry, to learn from it, and to use it for constructive actions" (p. 743).

Sometimes, the call to avoid provoking climate anxiety focuses in particular on young children. Strife (2012, p. 49) argues that young children "are not developmentally prepared to emotionally cope with global environmental issues"; while Sobel (1996, p. 4) warns that "if we prematurely ask children to deal with problems beyond their understanding and control, then … we cut them off from the possible sources of their strength." Both authors call for developmentally and age appropriate environmental education. In the second part of this claim, there is little to disagree with. A central principle of most approaches to social justice education (whether feminist, Freireian, anarchist, anti-racist, democratic, progressive, and so on) is to start with the experiences, needs, interests and capabilities of the learner. It is the first part of the claim made by Sobel and Strife that is contestable, for it draws on a longstanding western cultural image of children as innocent, fragile, vulnerable, and needing protection and shelter from the realities of the world (Moran-Ellis & Sünker, 2018; Taft, 2019). As discussed in Chap. 6, developing effective climate change education requires a substantive rethinking of this image of the child. Many children are already aware of the climate crisis, so denying them a "supportive space" to talk and think about this crisis can be more problematic than enabling them to do so (Kelsey & Armstrong, 2012, p. 190). "*Not* going public when the world is in turmoil may be much harder on children than for them to take up their share of the burden," writes Elshtain (1996, p. 20). Furthermore, there are numerous examples, from the climate crisis movement itself and other social and political movements in history, that suggest that many children are more than capable, not only of engaging with and understanding the climate crisis, but taking decisive and effective climate action (Coles, 1986; Taft, 2019). We are in the age now, after all, of the global school climate strike.

In recent years, there has been a social justice tradition in education that calls for educators to embrace a "pedagogy of discomfort," that "invites students to critique their deeply held assumptions, and to destabilize their views of themselves and their worlds" (Bozalek et al., 2014, p. 42). The assumption is that "discomforting feelings"—such as panic, fear, worry and anxiety—are "important in challenging dominant beliefs, social habits and normative practices," and can help to "create openings for individual and social transformation" (Zembylas, 2015, p. 163). Though the rhetoric is recent, there is a long history in radical education that parallels this notion of the value of student discomfort in education for social change. Myles Horton described the approach taken by the Highlander Folk School, centrally involved in the US civil rights movement, as based in a theory of "crisis education." "People's interest is heightened and they learn faster" when there is a "crisis situation," argues Horton (2003, p. 219), as they turn to education for "help with solving their own problems or in carrying out an urgently needed program."

Indeed, many and perhaps most social justice education traditions have developed out of a heightened sense of crisis—this was the case, for example, with the creation of Reggio Emilia in Northern Italy and Alex Bloom's democratic education experiments in East London in the immediate aftermath of the Second World War. Paulo Freire (2000), likewise, referred to his educational approach in *Pedagogy of the Oppressed* as "problem-posing education." Problem-posing education originally emerged in the context of adult literacy classes that Freire ran with poor agricultural workers in the northeast of Brazil in the late 1950s, in which Freire and other facilitators organised their classes around key "generative themes," that were "carefully chosen words that represented *the emotionally loaded and socially problematic issues* in participants' lives" (Wallerstein & Auerbach, 2004, p. 7; emphasis added). A pedagogy of discomfort is often framed in terms of teachers needing to actively "encourage students to move outside their 'comfort zones'" (Zembylas, 2015, p. 163); and, because of this framing, it has been critiqued as a "pedagogy for the privileged" (Walker & Palacios, 2016, p. 178). What Horton and Freire remind us is that discomfort, crisis and social problems often do not need to be introduced to students by educators. Rather, these frequently already exist in students' lives outside the classroom, and it is the task of educators to make sure that they reorient their educative work to support students in directly addressing the root causes of their pre-existing discomfort, their problems, their sense of impending crisis.

Paulo Freire, bell hooks and a Pedagogy of Hope

Fear and anxiety, just like hope, can potentially be both pivotal for and harmful to effective education and action for social change. This presents a dilemma for educators: if we can't just gamely seek to promote and embrace hope (or alternatively, fear and anxiety) in any certainty that this will offer a solid foundation for effective climate change education, then we need a way to find our bearings, to navigate and differentiate between good and bad, helpful and harmful forms of hope and/or fear. Action is often invoked here, as the key for turning climate hope or anxiety into something productive and effective (Pihkala, 2019). But, as noted earlier in the discussion of Ojala's concept of constructive hope, promotion of action raises a similar set of dilemmas. Action, too, can be good or bad, helpful or harmful (or something in between); even if we engage in copious amounts of climate change action, it could, as MacBride (2012) suggests, merely produce lots of "busy-ness" that doesn't actually accomplish much of substance, and worse, may distract valuable time and energy from core battles that need to be fought.

To develop a grounded understanding of exactly how and when hope and fear might be productive for progressive social transformation, it is helpful to turn to the concept of a pedagogy of hope, that is associated most closely with Paulo Freire and bell hooks, but that has close parallels in the various "utopian pedagogies" that have been developed by radical educators and activists over the course of the twentieth century (Coté et al., 2007). There are three core elements to a pedagogy of hope, and what is essential to note is that none of these elements focus principally on the individual, interior and psychological states of positivity or trust (in self, science or society) that have been a core concern for many current discussions of hope in relation to climate change education and communication. A pedagogy of hope, fundamentally, is a collective not just individual process, a social practice not a psychological orientation, and focuses outward on global social, political and economic structures, as much as it looks inward on habits of heart and mind. It is concerned with fostering what has been called a "critical" or "educated" form of hope (Bozalek et al., 2014; Freire, 1994; Webb, 2010).

At the heart of a pedagogy of hope is what hooks (2003, p. 195) calls "prophetic imagination." "Throughout my teaching career," hooks writes, "I have shared with students my belief in the power of prophetic imagination, telling them again and again 'that what we cannot imagine we cannot bring into being.'" Quoting the words of liberation theologian Mary Grey, hooks insists that prophetic imagination is not "some private day-dream, but is a fully public imagination, belonging to the public domain, inspiring the full range of communities belonging to it to commitment to fuller visions of well-being" (p. 195). Freire (1994, p. 77) writes similarly of "dreaming" as being "a mover of history" and "a necessary political act." "There is no change without dream, as there is no dream without hope," Freire insists:

> As project, as design for a different, less-ugly 'world', the dream is as necessary to political subjects, transformers of the world and not adapters to it, as … it is fundamental for an artisan, who projects in her or his brain what she or he is going to execute even before the execution thereof. (p.78)

Such sentiments have long been a cornerstone of radical education, to the degree that radical approaches to education are sometimes characterised as utopian pedagogies (Giroux, 2003). In the 1930s, George Counts' (1932, p. 37) influential pamphlet, *Dare the School Build a New Social Order?*, called on educators to "give to our children a vision of the possibilities which lie ahead and endeavour to enlist their loyalties and enthusiasms in the realization of the vision." Marxist scholar Erik Olin Wright (2003, p. vii), writing of the pursuit of "real utopias," argues just like hooks does that "what is pragmatically possible is not fixed independently of our imaginations, but is itself shaped by our visions." Judith Suissa (2010, p. 123) argues that "utopian hope" is at the heart of anarchist approaches to education for social change, which she describes as the "practice of imagining a world radically different from our own, and … daring to believe in its possibility," and the belief that "the very imaginative exercise of encouraging people to conceptualize the exact form of this [future] society, and to constantly engage with and experiment with its principles and manifestations, is itself a central part of the revolutionary process." Haiven and Khasnabish (2014, p. 3) point to the importance of the "radical imagination" for all social movement struggles, defining this as "the ability to imagine the world, life and social institutions not as they are but as they might otherwise be," and "the courage and intelligence to recognize that the world can and should be changed." Levitas (2013) describes "utopia as method," emphasizing the creative, intellectual labor that needs to be marshalled in order to develop a "serious utopianism," a process that entails "looking holistically at alternative modes of livelihood and social organisation, exposing them to public debate … and considering means of transition" (Levitas, 2008, p. 90; 2013).

But imagining, dreaming and hoping for a better, more utopian future is not enough, for there is always a danger that this becomes an escapist and superficial fantasy. "The idea that hope alone will transform the world," writes Freire (1994, p. 2), "and action undertaken in that kind of naïveté, is an excellent route to hopelessness, pessimism, and fatalism. … [T]here is no hope in sheer hopefulness." Prophetic imagination needs to be combined with comprehensive, critical analysis of the current "situation of oppression," and how a better future may be built out of the sobering reality of where we exist today (Freire, 1994, p. 23). Freire thus argues that "one of the tasks of the progressive educator, *through a serious, correct political analysis*, is to unveil opportunities for hope, no matter what the obstacles may be" (p. 3, emphasis added). As Papastephanou (2016, pp. 32, 44) notes, Freire promotes not just a "pedagogy of hope" but a "pedagogy of fear" as well, one that pushes teachers and students "to detect gloomy realities, to voice discontent, and to cultivate dystopian criticality." In Freire's (1985) terminology, effective utopian education requires "the dramatic unity of denunciation and annunciation." "Denunciation of a dehumanizing situation … demands precise scientific understanding of that situation," writes Freire (1985), while "the annunciation of its transformation … requires a theory of transforming action" as well as a guiding or "authentic utopian vision."

Once again, similar arguments may be found across different traditions of twentieth century radical education. Effective, critical hope, for liberation theologian Gustavo Gutiérrez (1973, p. 238) needs to be driven by a "liberating utopia" as well as a "scientific analysis of reality."

For hooks (2003, p. xiii, xiv), prophetic imagination should be combined with "rigorous critique," in order to "understand better the ways diverse systems of domination operate;" as "our visions for tomorrow are most vital when they emerge from the concrete circumstances of change we are experiencing right now." For Counts (1932, p. 2), educators in the United States who want to contribute effectively to radical social change must foster their utopian imagination, but also "must abandon much of their easy optimism, subject the concept of education to the most rigorous scrutiny, and be prepared to deal much more fundamentally, realistically and positively with the American social situation than has been their habit in the past." In Wright's (2003, p. vii) account, "real utopias" need to "be grounded in the real potentials of humanity," "accommodate to practical realities," provide "pragmatically accessible waystations," and respond to what is always "a world of imperfect conditions for social change."

Finally, a pedagogy of hope needs to mobilize imagination and critique, annunciation and denunciation, to inform direct, collective strategies for social change. As Freire (1994, p. 3) writes at the beginning of *Pedagogy of Hope*:

> Without a minimum of hope, we cannot so much as start the struggle. But without the struggle, hope … dissipates, loses its bearings, and turns into hopelessness. … Hopelessness and despair are both the consequence and cause of inaction or immobilism.

Freire's (2000, pp. 87, 88) vision of liberatory pedagogical praxis is one that insists on the close linking of education and action, to avoid the perils of both "verbalism" (or "idle chatter") and "activism" (or "action for action's sake"). The argument of hooks in *Teaching Community: A Pedagogy of Hope* is closely parallel: one of the tasks of a pedagogy of hope, writes hooks (2003, pp. xvi, 12, 197), is to foster "communities of resistance" that are "life-enhancing" and "life-sustaining," that are brought into being by collective, critical hope, and that work to restore and renew such hope. Again, the vital links between critique, imagination and collective action have long been emphasized across different traditions of radical education (Tarlau, 2014). "Education without social action … has no true power potential," Martin Luther King Jr. (1968, p. 164) famously warned, while "social action" that is "uninformed by educated thought can take false directions" and amount to little more than "a weak statement of pure energy."

If we use the example of plastics recycling discussed earlier, then a pedagogy of hope would be based not simply in a positive, hopeful, enthusiastic rush to embrace local recycling initiatives, but in careful, sustained, thorough and rigorous analysis of the current realities of, among other things, both plastics production and the global recycling industry. It would focus our attention on asking exactly how and why corporate plastic producers have been so keen over the years to promote individual consumer responsibility for the recycling of plastic; or how and why so little plastic is ever recycled, despite decades of promoting recycling practice; or what the impact on communities and environments overseas actually are from the plastic that we so carefully and enthusiastically recycle. It would also focus on developing collective actions that might have real hope of addressing the problem of plastic waste pollution, and thereby helping to create a better, more sustainable world in the future—perhaps, for example, through targeting directly corporate plastic production in the first place. How could such a pedagogy of hope be developed and judged? It is certainly not a matter of how well it vanquishes student fear or anxiety, nor any compulsion to positivity that it might evoke. Rather, it needs to be assessed on the strength and comprehensiveness of its critique, the clarity and consistency of an inspiring vision for the future, and the pragmatism and strategy of putting forth a workable theory and plan for the carrying out of action that might help get us to there from here.

Emotions and Educating for Radical Social Change

Sarah Amsler (2011) points out that many and probably most progressive and radical education traditions embrace emotions as a vital part of the holistic approaches to teaching and learning they consider essential to facilitating social transformation. At the same time, a strong critique has emerged about the "affective turn" in education over the past few years (that includes the push to promote positivity, hope, well-being, self-esteem and happiness among students), that is sometimes "derided as individualistic 'therapeutic pedagogy', counter to transformational change" (Walker & Palacios, 2016, p. 175). There is, Amsler notes, a tension or contradiction that needs to be worked through here. "Clearly, affect is central to both learning and to any viable conception of socially responsive education," writes Amsler (2011, p. 52), so "how might we avoid both incorporation into 'therapeutic' or consumerist practices of education, on the one hand, and the exclusion of affect itself, on the other?" Part of the resolution to this contradiction, Amsler suggests, may be found in the pedagogy of hope discussed above, that ties emotions such as hope and fear directly to careful critical analysis of social structures and collective, strategic projects of social, utopian action. But there remains the question of emotions in education, in and of themselves.

As the issue of emotion attracts growing attention in climate change education, there are two claims that can be more or less universally supported. One is that learning, and especially learning for social change, entails emotional engagement and transformation, and is rarely just a cognitive undertaking (Russell & Oakley, 2016; Walker & Palacios, 2016). "Our emotions help us to envision future horizons of possibilities and who we want to become," writes Boler (1999, p. xv), "they give us information about what we care about and why," and "shape the selectivity of our cognitive and ethical attention and vision." The second is that the climate crisis itself, as well as the challenge of taking effective action to address the crisis, often produce strong emotions for learners, and these emotions can have powerful effects for individuals and communities (Dietz et al., 2020; Verlie, 2019). Beyond these two claims, it is difficult to make many further general arguments about the relationships between emotions and climate change education and action: not everyone experiences strong emotions or the same emotions in relation to the global climate crisis; emotions, such as hope and fear, do not always have the same effects on individual and collective action (or inaction); and the outcomes from different emotions in response to climate crisis may shift over time, so that short-term responses to climate fear and anxiety, might be quite different to longer-term consequences, as initial depression and paralysis develops into extended engagement and action (or vice versa) (Chapman et al., 2017; Salama & Aboukoura, 2018).

It is within feminist pedagogical traditions that probably the most sustained work has been done in thinking about the significance and place of emotions in radical education projects for social change (Boler, 1999; Sayles-Hannon, 2007). Through a critical engagement with this literature, several key principles might be suggested. Educators should work to avoid promoting or labelling different kinds of emotions as being better or worse, wrong or right, preferred or dispreferred in relation to climate change education and action. Not only is there no clear basis for making such claims, but once we enter into the process of such labelling, we shift education into becoming "a site for social control," and risk shutting down spaces for learning for at least some of our students (Boler, 1999, p. xiv; Walker & Palacios, 2016). Climate change educators should also avoid taking an instrumental approach to emotions, seeking to create and use certain emotions to produce desired effects. As Sara Ahmed (2004, pp. 181–182) observes:

> [E]motions are critical to feminist pedagogy. … [But] to make this claim is not to idealise emotions as 'good' or necessary to critical thinking or learning. Indeed, emotions should not become the preferred 'outcome' of teaching. This would transform emotions into a bank, to evoke Freire's … classic metaphor for instrumental and conservative practices of teaching. If emotions were to become the 'outcome' of feminist teaching (rather than part of the process), then the task of the feminist teacher would be to "fill' the students with' the right emotions … Here, emotions would be transformed into fetish objects, which we assumed had meaning in advance of their naming.

There is no basis for thinking that "appeals to specific emotions will be associated with specific outcomes," or that emotions are "magic bullets" or "unitary switches that induce specific behavioural responses" (Chapman et al., 2017, pp. 850, 851). Finally, educators should approach emotions as being absolutely central to the learning process, whether about the climate crisis or any other topic, while accepting that the emotionality of this learning process might work differently for different individuals in different contexts. "Emotions should be acknowledged, honoured and interrogated, not promoted or manipulated" in climate change education classrooms, argues Verlie (2019, p. 752); likewise, Hufnagel (2017, p. 52) call for educators to be "attending to emotions in climate change education," but not prescribing certain preferred emotions over others. In this, there are links with the liberatory or emancipatory approach of the pedagogy of hope developed by Freire, hooks and others. "I have a right to be angry, to show it and to use it as a motivational foundation for my struggle," Freire (1998, p. 71) writes in *Pedagogy of Freedom*, "just as I have a right to love and to express my love to the world and to use it as a motivational foundation for my struggle."

References

Ahmed, S. (2004). *The cultural politics of emotion*. Edinburgh University Press.

Amsler, S. (2011). From 'therapeutic' to political education: The centrality of affective sensibility in critical pedagogy. *Critical Studies in Education, 52*(1), 47–63.

Autere, S. (2019). Sustainable lifestyles help with climate anxiety. *Sitra*, 27 November. https://www.sitra.fi/en/news/sustainable-lifestyles-help-with-climate-anxiety/

Bilodeau, C. (Ed.). (2018). *Where is the hope? An anthology of short climate change plays*. Climate Change Theatre Action.

Binkley, S. (2014). *Happiness as enterprise*. SUNY Press.

Bloomberg, M., & Pope, C. (2017). *Climate of hope*. St Martin's Press.

Blum, D. (2020). How climate anxiety is shaping family planning. *New York Times*, 15 April.

Boler, M. (1999). *Feeling power: Emotions and education*. Routledge.

Boyce, G. (2014). Hope scale. In A. Michalos (Ed.), *Encyclopedia of quality of life and well-being research* (pp. 2925–2926). Springer.

Boyte, H. (2004). *Everyday politics: Reconnecting citizens and public life*. University of Pennsylvania Press.

Bozalek, V., Carolissen, R., & Leibowitz, B. (2014). A pedagogy of critical hope in South African higher education. In V. Bozalek et al. (Eds.), *Discerning critical hope in educational practices* (pp. 40–54). Routledge.

Bretherton, L. (2015). *Resurrecting democracy: Faith, citizenship, and the politics of a common life*. Cambridge University Press.

Brown, J. (2017). Is positive thinking the way to save the planet? *New Scientist*, 11 October. https://www.newscientist.com/article/mg23631470-800-is-positive-thinking-the-way-to-save-the-planet/

Bryan, A. (2020). Affective pedagogies: Foregrounding emotion in climate change education. *Policy and Practice: A Development Education Review, 30*, 8–30.

Cabanas, E., & Illouz, E. (2019). *Manufacturing happy citizens: How the science and industry of happiness control our lives*. Polity Press.

Cambridge Network. (2020). Sir David Attenborough takes a stand for Earth Optimism. 3 March. https://www.cambridgenetwork.co.uk/news/sir-david-attenborough-takes-stand-earth-optimism

Chapman, D., Lickel, B., & Markowitz, E. (2017). Reassessing emotion in climate change communication. *Nature Climate Change, 7*, 848–852.

Coles, R. (1986). *The political life of children*. Atlantic Monthly Press.

Coté, M., Day, R., & de Peuter, G. (2007). Utopian pedagogy: Creating radical alternatives in the neoliberal age. *Review of Education, Pedagogy and Cultural Studies, 29*, 317–336.

Counts, G. (1932). *Dare the school build a new social order?* Southern Illinois University Press.

Davies, W. (2015). *The happiness industry: How the government and big business sold us well-being*. Verso.

Devex. (2020). Global Optimism. https://www.devex.com/organizations/global-optimism-135686

Dietz, T., Shwom, R., & Whitley, C. (2020). Climate change and society. *Annual Review of Sociology, 46*(5), 1–24.

Doran, C. (2017). *Hope in the age of climate change*. Cascade Books.

Earth Optimism. (2020a). About Earth Optimism. http://www.earthoptimism.cambridgeconservation.org/about-earth-optimism/

Earth Optimism. (2020b). Special message from the Alliance. http://www.earthoptimism.cambridgeconservation.org/special-message/

Ehrenreich, B. (2009). *Bright-sided: How the relentless promotion of positive thinking has undermined America*. Metropolitan Books.

Elshtain, J. (1996). Commentary: Political children. *Childhood, 3*, 11–28.

Euronews. (2020). Davos 2020: Donald Trump and Greta Thunberg clash on climate at World Economic Forum. 22 January. https://www.euronews.com/2020/01/21/greta-thunberg-tells-davos-leaders-her-climate-movement-is-just-the-beginning

Figueres, C., & Rivett-Carnac, T. (2020). *The future we choose: Surviving the climate crisis*. Alfred A. Knopf.
Flannery, T. (2015). *Atmosphere of hope*. Harper Collins.
Freire, P. (1985). *The politics of education: Culture, power & liberation*. Bergin & Garvey.
Freire, P. (1994). *Pedagogy of hope*. Bloomsbury.
Freire, P. (1998). *Pedagogy of freedom*. Rowman & Littlefield.
Freire, P. (2000). *Pedagogy of the oppressed*. Continuum.
Fridays for Future. (2020). *Our house is on fire*. https://www.youtube.com/watch?v=eT32UFzA7E8
Ghosh, A. (2016). *The great derangement*. University of Chicago Press.
Giroux, H. (2003). Utopian thinking under the sign of neoliberalism: Towards a critical pedagogy of educated hope. *Democracy & Nature, 9*(1), 91–105.
Global Action Plan. (2020). *Climate anxiety: An introduction for teachers*. Global Action Plan.
Global Optimism. (2020). Global Optimism. https://globaloptimism.com
González-Gaudiano, E., & Meira-Cartea, P. (2019). Environmental education under seige: Climate radicality. *Journal of Environmental Education, 50*(4), 386–402.
Gosden, E. (2014). Look on bright side of climate change, says IPCC report author. *Telegraph*, 31 March. https://www.telegraph.co.uk/news/earth/environment/climatechange/10733498/Look-on-bright-side-of-climate-change-says-IPCC-report-author.html
Gutiérrez, G. (1973). *A theology of liberation*. Orbis.
Haiven, M., & Khasnabish, A. (2014). *The radical imagination*. Zed Books.
Haltinner, K., & Sarathchandra, D. (2018). Climate change skepticism as a psychological coping strategy. *Sociology Compass, 12*(e12586), 1–10.
Hammonds, B. (2008). Environmental education or ecophobia! *Leading and Learning*, 4 June. https://leading-learning.blogspot.com/2008/06/environmental-education-or-ecophobia.html
Head, L. (2016). *Hope and grief in the Anthropocene*. Routledge.
Held, B. (2002). The tyranny of the positive attitude in America. *Journal of Clinical Psychology, 58*(9), 965–991.
Hickman, C. (2019). What psychotherapy can do for the climate and biodiversity crises. *The Conversation*, 7 June. https://theconversation.com/what-psychotherapy-can-do-for-the-climate-and-biodiversity-crises-116977
Hilliard, M. (2020). Cork primary school's viral rap video wins European award. *Irish Times*, 15 May. https://www.irishtimes.com/news/ireland/irish-news/cork-primary-school-s-viral-rap-video-wins-european-award-1.4254742

hooks, b. (2003). *Teaching community: A pedagogy of hope*. Routledge.
Horowitz, D. (2018). *Happier? The history of a cultural movement that aspired to transform America*. Oxford University Press.
Horton, M. (2003). Crisis education. In D. Jacobs (Ed.), *The Myles Horton reader* (pp. 219–220). University of Tennessee Press.
Hufnagel, E. (2017). Attending to emotional expressions about climate change: A framework for teaching and learning. In D. Shepardson (Ed.), *Teaching and learning about climate change: A framework for educators* (pp. 59–71). Routledge.
Ideland, M. (2015). The end of the world and a promise of happiness: Environmental education within the cultural politics of emotions. In T. Popkewitz, J. Diaz, & C. Kirchgasler (Eds.), *A political sociology of educational knowledge*. Routledge.
Irish Times. (2020). One small change: Tiny Irish school's climate-crisis rap going viral. 8 February. https://www.irishtimes.com/culture/tv-radio-web/one-small-change-tiny-irish-school-s-climate-crisis-rap-going-viral-1.4166788
Jorgenson, S., Stephens, J., & White, B. (2019). Environmental education in transition: A critical review of recent research on climate change and energy education. *Journal of Environmental Education, 50*(3), 160–171.
Kelsey, E. (2016). Propagating hope in the midst of environmental doom and gloom. *Canadian Journal of Environmental Education, 21*, 23–40.
Kelsey, E., & Armstrong, C. (2012). Finding hope in a world of environmental catastrophe. In A. Wals & P. Corcoran (Eds.), *Learning for sustainability in times of accelerating change* (pp. 187–200). Wageningen Academic Publishers.
Kidd, L., Bekessy, S., & Garrard, G. (2019). Neither hope nor fear: Empirical evidence should drive biodiversity conservation strategies. *Trends in Ecology & Evolution, 34*(4), 278–282.
King, M. L. K., Jr. (1968). *Where do we go from here: Chaos or community?* Beacon Press.
Klein, N. (2014). *This changes everything*. Penguin.
Kundzewicz, Z., et al. (2020). From 'atmosfear' to climate action. *Environmental Science and Policy, 105*, 75–83.
Leiserowitz, A., et al. (2018). *Climate change in the American mind*. Yale Program on Climate Change Communication.
Lerner, S. (2019a). Africa's exploding plastic nightmare. *The Intercept*, 19 April. https://theintercept.com/2020/04/19/africa-plastic-waste-kenya-ethiopia/

Lerner, S. (2019b). Waste only: How the plastics industry is fighting to keep polluting the world. *The Intercept*, 20 July. https://theintercept.com/2019/07/20/plastics-industry-plastic-recycling/
Levitas, R. (2008). Be realistic: Demand the impossible. *New Formations, 65*, 78–93.
Levitas, R. (2013). *Utopia as method*. Palgrave Macmillan.
Li, C., & Monroe, M. (2018). Development and validation of the Climate Change Hope Scale for high school students. *Environment and Behavior, 50*(4), 454–479.
MacBride, S. (2012). *Recycling reconsidered: The present failure and future promise of environmental action in the United States*. MIT Press.
Maniates, M. (2001). Individualization: Plant a tree, buy a bike, save the world? *Global Environmental Politics, 1*(3), 31–52.
Maniates, M. (2016). Make way for hope: A contrarian view. In S. Nicholson & S. Jinnah (Eds.), *New earth politics: Essays from the Anthropocene* (pp. 135–154). MIT Press.
Mann, M., & Toles, T. (2016). *The madhouse effect*. Columbia University Press.
Mission2020. (2020). Mission2020. https://mission2020.global
Moltman, J. (2019). *The spirit of hope: Theology for a world in peril*. Westminster John Knox Press.
Moran-Ellis, J., & Sünker, H. (2018). Childhood studies, children's politics and participation: Perspectives for processes of democratisation. *International Review of Sociology, 28*(2), 277–297.
Norgaard, K. (2011). *Living in denial: Climate change, emotions, and everyday life*. MIT Press.
Ojala, M. (2007). Confronting macrosocial worries: Worry about environmental problems and proactive coping among a group of young volunteers. *Futures, 39*, 729–745.
Ojala, M. (2008). Recycling and ambivalence: Quantitative and qualitative analyses of household recycling among young adults. *Environment and Behavior, 40*(6), 777–797.
Ojala, M. (2012a). Hope and climate change: The importance of hope for environmental engagement among young people. *Environmental Education Research, 18*(5), 625–642.
Ojala, M. (2012b). Regulating worry, promoting hope: How do children, adolescents, and young adults cope with climate change? *International Journal of Environmental & Science Education, 7*(4), 537–561.
Ojala, M. (2015). Hope in the face of climate change: Associations with environmental engagement and student perceptions of teachers' emotion

communication style and future orientation. *Journal of Environmental Education, 46*(3), 133–148.
Ojala, M. (2016). Facing anxiety in climate change education: From therapeutic practice to hopeful transgressive learning. *Canadian Journal of Environmental Education, 21*, 41–56.
Ojala, M. (2017). Hope and anticipation for a sustainable future. *Futures, 94*, 76–84.
Papastephanou, M. (2016). Needing 'tomorrow as fish need water': Dystopia, utopia, and Freire's pedagogy. *Interchange, 47*, 31–49.
Pihkala, P. (2017). Environmental education after sustainability: Hope in the midst of tragedy. *Global Discourse, 7*(1), 109–127.
Pihkala, P. (2019). *Climate anxiety*. MIELI Mental Health Finland.
Plautz, J. (2020). Eco-anxiety is overwhelming kids: Where's the line between education and alarmism? *Washington Post*, 3 February.
Raygorodetsky, G. (2017). *The archipelago of hope*. Pegasus.
Reser, J., & Bradley, G. (2017). Fear appeals in climate change communication. In M. Nisbet et al. (Eds.), *The Oxford encyclopedia of climate change communication*. Oxford University Press.
Russell, C., & Oakley, J. (2016). Engaging the emotional dimensions of environmental education. *Canadian Journal of Environmental Education, 21*, 13–22.
Salama, S., & Aboukoura, K. (2018). Role of emotions in climate change communication. In W. Filho et al. (Eds.), *Handbook of climate change communication: Vol. 1* (pp. 137–150). Springer.
Sandford, R., & O'Riordan, J. (2017). *The hard work of hope*. Rocky Mountain Books.
Saylan, C., & Blumstein, D. (2011). *The failure of environmental education*. University of California Press.
Sayles-Hannon, S. (2007). Feminist and liberatory pedagogies: Journey toward synthesis. *International Journal of Diversity in Organisations, Communities & Nations, 7*(2), 33–42.
Scicluna, G. (2019). Greta Thunberg, the prophet of doom. *MaltaWinds*, 26 September. http://maltawinds.com/2019/09/26/greta-thunberg-the-prophet-of-doom/
Selby, D., & Kagawa, F. (2018). Teetering on the brink: Subversive and restorative learning in times of climate turmoil and disaster. *Journal of Transformative Education, 16*(4), 302–322.

Snyder, C., Rand, K., & Sigmon, D. (2005). Hope theory: A member of the positive psychology family. In C. Snyder & S. Lopez (Eds.), *Handbook of positive psychology* (pp. 257–276). Oxford University Press.
Soave, R. (2019). Here's what's wrong with *Time* declaring Greta Thunberg Person of the Year. *Reason*, 11 December. https://reason.com/2019/12/11/greta-thunberg-time-person-of-the-year/
Sobel, D. (1996). *Beyond ecophobia*. Orion Society.
Sobel, D. (2007). Climate change meets ecophobia. *Connect, 21*(2), 14–21.
Stevenson, K., & Peterson, N. (2015). Motivating action through fostering climate change hope and concern and avoiding despair among adolescents. *Sustainability, 8*(6), 1–10.
Stevenson, R., Nicholls, J., & Whitehouse, H. (2017). What is climate change education? *Curriculum Perspectives, 37*, 67–71.
Strife, S. (2012). Children's environmental concerns: Expressing ecophobia. *Journal of Environmental Education, 43*(1), 37–54.
Suissa, J. (2010). *Anarchism and education: A philosophical perspective*. PM Press.
Sutton, K., & Smith, L. (2020). Ecoanxiety will help us save the planet. *Nesta*. https://www.nesta.org.uk/feature/ten-predictions-2020/ecoanxiety-will-help-us-save-planet/?gclid=EAIaIQobChMI9-6S5df06QIVhrTtCh30HQeHEAAYASAAEgKZ4fD_BwE
Svoboda, M. (2020). New and recent books about hope in a time of climate change. Yale Climate Connections, 21 February. https://www.yaleclimateconnections.org/2020/02/new-and-recent-books-about-hope-in-a-time-of-climate-change/
Taft, J. (2019). *The kids are in charge: Activism and power in Peru's movement of working children*. New York University Press.
Tarlau, R. (2014). From a language to a theory of resistance: Critical pedagogy, the limits of 'framing' and social change. *Educational Theory, 64*(4), 369–392.
Thucydides. (1974). *History of the Peloponnesian War*. Trans. R. Warner. Penguin.
Thunberg, G. (2019a). Our house is on fire. *Guardian*, 25 January. https://www.theguardian.com/environment/2019/jan/25/our-house-is-on-fire-greta-thunberg16-urges-leaders-to-act-on-climate
Thunberg, G. (2019b). Cathedral thinking. 16 April. In *No one is too small to make a difference*. Penguin.
Thunberg, G. (2020). Our house is still on fire. *Democracy Now!* 21 January. https://www.democracynow.org/2020/1/21/our_house_is_still_on_fire
Tucci, J., Mitchell, J., & Goddard, C. (2007). *Children's fears, hopes and heroes: Modern childhood in Australia*. Australian Childhood Foundation.

Verlie, B. (2019). Bearing worlds: Learning to live with climate change. *Environmental Education Research, 25*(5), 751–766.
Verplanken, B., & Roy, D. (2013). 'My worries are rational, climate change is not': Habitual ecological worrying is an adaptive response. *PLoS One, 8*(9), 1–6.
Waldron, F., Ruane, B., Oberman, R., & Morris, S. (2019). Geographical process or global injustice? Contrasting educational perspectives on climate change. *Environmental Education Research, 25*(6), 895–911.
Walker, J., & Palacios, C. (2016). A pedagogy of emotion in teaching about social movement learning. *Teaching in Higher Education, 21*(2), 175–190.
Wallace-Wells, D. (2019). *The uninhabitable earth*. Penguin.
Wallerstein, N., & Auerbach, E. (2004). *Problem-posing at work: Popular educator's guide*. Grass Roots Press.
Walls, D. (2015). *Community organizing: Fanning the flames of democracy.* Polity Press.
Ward, M. (2019). Climate anxiety is real, and young people are feeling it. *Sydney Morning Herald*, 20 September. https://www.smh.com.au/lifestyle/health-and-wellness/climate-anxiety-is-real-and-young-people-are-feeling-it-20190918-p52soj.html
Webb, D. (2010). Paulo Freire and the 'the need for a kind of education in hope'. *Cambridge Journal of Education, 40*(4), 327–339.
Webb, D. (2013). Pedagogies of hope. *Studies in the Philosophy of Education, 32*, 397–414.
Whitmore-Williams, S., et al. (2017). *Mental health and our changing climate.* American Psychological Association.
Wright, E. (2003). The real utopias project. In A. Fung & E. Wright (Eds.), *Deepening democracy*. Verso.
Yang, Z., & Kahlor, L. (2012). What me worry? The role of affect in information seeking and avoidance. *Science Communication, 35*(2), 189–212.
Young, S. (2020). Eco-anxiety: Children are losing sleep and having bad dreams over climate change, study finds. *Independent*, 3 March. https://www.independent.co.uk/life-style/children-climate-change-sleep-nightmares-eco-anxiety-greta-thunberg-a9371191.html#gsc.tab=0
Zembylas, M. (2015). 'Pedagogy of discomfort' and its ethical implications: The tensions of ethical violence in social justice education. *Ethics and Education, 10*(2), 163–174.

5

Thinking Through Places

Formal education has often been associated with a process of deterritorialization, disembedding or taking out of place. At its most immediate level, this is experienced by students as a leaving behind of home, family and community to continue with schooling, particularly at the secondary and post-secondary levels. "The connections between a movement away from place and the processes of formal education are so well established and engrained that we seldom even recognize them," writes Michael Corbett (2009, p. 2). But disembedding from place is woven into the fabric of formal education well beyond this. The development of learning has tended to be constructed in hierarchies that seek to move the learner from embeddedness in the concrete and local to ostensibly higher order capabilities of abstract, universalist thinking. Nation-state schooling systems seek to impose uniform national curricula that all children learn, irrespective of local place, dialect or community (Green, 1990); and imperialist, colonialist models of education have regularly inducted children into modes and subjects of learning far removed from their own communities. As Edward Said (2000, p. 392) observes of the Arab world under colonialism, education was "dominated by foreign rulers, … who taught us to respect distant norms and values more than our own,"

S. Tannock, *Educating for Radical Social Transformation in the Climate Crisis*, Palgrave Studies in Education and the Environment,
https://doi.org/10.1007/978-3-030-83000-7_5

dismissing "our culture" as "lower grade, perhaps even congenitally inferior and something of which to be ashamed." In today's global capitalist economy, policymakers call for education to be "aimed at producing young citizens who can function effectively anywhere in the postindustrial world," through their acquisition of allegedly "global" knowledge and skill (Noddings, 2005, p. 57). So linked is modern, western schooling with placelessness that the IPCC's (2018, pp. 552, 553) *Global Warming of 1.5 °C* report, adopting terminology from the United Nations, uses the phrase "Indigenous and local knowledge" as its counterpoint, defined as what mainstream schooling is not: "Indigenous knowledge" refers to "understandings, skills and philosophies developed by societies with long histories of interaction with their natural surroundings," and "local knowledge" includes "understandings and skills developed by individuals and populations, specific to the places where they live."

For environmental education, by contrast, place has long been embraced as pivotal to developing an effective education that can address environmental degradation and crisis, and environmental educators have sought to reinsert place into the educational process and experience. David Orr (1992, pp. 126–131) argues that "place has no particular standing in contemporary education," and modern schooling has helped to fashion us as a "deplaced people," whose "sources of food, water, livelihood, energy, [and] materials" come from "places around the world that are largely unknown to us, as are those to which we consign our toxic and radioactive wastes, garbage, sewage, and industrial trash." For Orr, "integration of place into education" is essential for addressing local and global problems of environmental devastation, by enabling us to relearn "the art of living well" in "our own places." The dictionary definition of "environment" as "the surroundings or conditions in which a person, animal or plant lives or operates," suggests that all forms of environmental education will have a central concern with "place," which is commonly understood as being "a meaningful location" (Stevenson, 2015; Cresswell, 2015, p. 13). Scholars such as David Sobel, David Gruenewald and Gregory Smith use this insight to argue for a model of "place-based" or "place-conscious" education as a framework for all (environmental) education (Smith, 2020).

The core idea of why place is important for environmental education, and climate change education more specifically, is relatively straightforward. It is argued that fostering or connecting with strong attachments to place can be an important motivator for taking climate change action, whether place is conceived of as a specific locality, region, nation or the planet itself; and that developing a grounded, experiential knowledge of local places and the global ties that link different local places together can be vital for understanding the impacts of human activity on the environment and learning to live sustainably within places, across different scales of the local, the national and the global (Devine-Wright, 2013; Khadka et al., 2021; Kudryavtsev et al., 2012; Schweizer et al., 2013). In the context of the climate crisis, a focus on place is said to make the overwhelming scale of the crisis more comprehensible, meaningful and actionable. "Effective climate education is … notoriously difficult because concepts such as the global climate system and long-term climate change often appear abstract, invisible, nebulous, technical, and unrelatable," writes Peter Rudiak-Gould (2014, p. 1): a focus on place can help teach about the crisis "by linking it to specific local places where the effects of climatic fluctuations are visible, and which carry aesthetic, spiritual, sociocultural, or practical value for local communities." Though a global phenomenon, the climate crisis is inevitably situated in places, as the human activities that create and exacerbate the crisis, the frequently devastating impacts of the crisis, and the collective actions needed to address the crisis, always occur in particular ways in specific places around the world (Hulme, 2008).

However, if there is a broad consensus on the importance of place in developing effective environmental and climate change education, there is considerable disagreement on exactly how place is important, and how place should be addressed and articulated within educational practice. To develop a critical and effective engagement with place in climate change education, therefore, there is a need not just to understand the arguments for why a focus on place in education is important and useful; but also to think through the different critiques of invocations of place in educational theory and practice. Place can be pivotal in climate change education, but only when considered within broader political, historical and global geographical contexts, and when relationships of interdependence and reciprocity that exist between human practice and experience, and nature and the non-human, are carefully and critically interrogated.

The Centrality of Place in Environmental Education

Place is a quintessential component of much if not all environmental education. Search for environmental education images on the internet and you will find pictures of children planting trees, working in school gardens, going on field trips in the woods, collecting water samples from ponds, and so on. Sobel's (2004, p. 1) evocative depiction of place-based education as a school that has been emptied of children, who are instead outside learning in the town woodlot, by an urban stream and in a schoolyard garden, holds true for much environmental education as well. Place matters well beyond the educational traditions that explicitly invoke the term in their self-definitions: place-based education, place conscious education, place responsive education, critical pedagogy of place, or radical pedagogy of place (Malone, 2017; Stevenson et al., 2018). As suggested above, place is likely to be central to any educational agenda concerned with the environment, given the considerable overlap between the two concepts. But beyond this, there are two key ideas about what makes for effective education that drive the concern with place in much environmental education, both of which are closely linked with the long traditions of progressive and radical education.

One is the idea that learning is most effective when it begins with the contexts, lives and interests of the learner, and when it is seen as being directly, personally meaningful. As Martha Monroe and her colleagues (Monroe et al., 2019, p. 799) found in a systematic research review, effective climate change education strategies make "climate change information *personally relevant and meaningful* for learners" (emphasis in original). If we seek to teach about the environment, it is often helpful to begin with the local environment immediately around us; and if we seek to teach about the climate crisis, it is helpful to start with how the crisis is directly relevant for the places and communities in which our students live and work. In the context of the climate crisis, this learner centered approach is a matter of attending to the challenges of scale, by making a globally diffuse phenomenon locally and concretely meaningful, but also of distance. Focusing on popular climate crisis emblems of vanishing rainforests and endangered polar bears, as important and striking as these phenomena may be, might not always be the best educational strategy for learners who live in regions where there are no polar bears or rainforests (Burningham & Thrush, 2001). "When the polar bear is the most visible mascot of climate change," argues Katharine Hayhoe (2017), "it does the rest of us a disservice by making the issue seem remote and distant." A second core idea is that learning is most effective when it is embodied, holistic and experiential, not just confined to a lecture, classroom or textbook, but built through the direct engagement of head, heart and hands all together. If we want to understand our local environments, we need to go out into these environments directly, to observe, measure, analyse, inquire, reflect and take action. School gardens, forest schools, outdoor education, conservation fieldtrips all come out of a broad tradition of experiential education that embraces the importance of hands on "learning by doing" (Roberts, 2012). Again, not coincidentally, Monroe et al.'s (2019, p. 799) systematic review of climate change education research finds that active, engaged, experiential and inquiry-based approaches constitute the second of two core characteristics shared by most effective climate change education strategies.

All of this "learning in places" can undoubtedly provide for engaging, fun, powerful forms of environmental education. But the question of what any of this does directly to help address the climate crisis, or other environmental crises, is less certain. One issue is that place, in itself, does not determine meaning or ideology or agenda—something that risks being misrecognized when claims are made by educators that "places are powerfully pedagogical," that "places themselves have something to say," or that we need "to listen to what places are telling us" (Greenwood, 2009, p. 275; Gruenewald, 2003, pp. 624, 645). Taking students to a forest, a watershed, a coastline, a toxic waste site or a school garden does not predetermine what teachers will teach in any of these sites, nor what any of these students will learn (Nakagawa & Payne, 2015). Take the example of school and community gardens, which have become increasingly popular in western countries, and are often enthusiastically promoted as a key site for climate change education (BC Farms & Food, 2019; RHS, 2019; Wake & Birdsall, 2016). As educator Jeannie Russell (2019) writes:

> Teaching about climate issues connected to food, farms, and global food production systems offers us countless ways of directly engaging our students, and highlights one of the most powerful contributors to both climate and overall environmental destabilization. … [T]he most engaging and beneficial component of climate education around food and farms [are] community and school gardens. … School and community gardens are places where education and activism around climate change and habitat degradation blend into an engaging and inspiring mix: teaching about and actively supporting sustainable small-scale farming methods, bringing community together to tackle issues of food insecurity and food deserts, while making neighborhood habitats more diverse and resilient.

Yet, school gardens have existed throughout much of the history of modern schooling, and in different times and places, have been used to support a range of different ideologies and agendas, many of which are not particularly helpful for addressing the challenge of the climate crisis: from celebrating militarism and patriotism (during the First and Second World Wars), to supporting elite agendas of social control of immigrants, youth and the poor, to "promoting a neoliberal ethos of individual responsibility for systemic problems" (Cairns, 2018, p. 517; see also Mah, 2010; Pudup, 2008; Trelstad, 1997). Places, whether tiny school gardens or massive continental watersheds, are socially constructed and have multiple, conflicting and contested meanings (Cresswell, 2015). A forest, as Claudia Ruitenberg (2005, pp. 218–219) points out, "is a site of economic benefit to the logging and tourism industry, as well as an ecosystem, as well as land formerly inhabited by Indigenous people."

In environmental education, it is sometimes assumed that fostering or connecting with strong attachments to place will lead directly to greater environmental care and action among students. "When people acquire a deep knowledge of a particular place, they begin to care about what happens to the landscape, creatures, and people in it," writes Michael Stone (2009, p. 13): "Places known deeply are deeply loved, and well-loved places have the best chance to be protected and preserved, to be cherished and cared for by future generations." But while this can and often does happen, it is not inevitably the case, for it depends on the kinds of meanings and attachments that places have for different people (Kudryavtsev et al., 2012). As David Chang (2017, p. 727) observes:

> One's love of place does not necessarily compel one to protect place. In Canada, ... many people enjoy outdoor recreational activities, such as skiing, camping, water-sports and hunting. Some of these activities require a great deal of equipment, as well as capacious, fuel-guzzling vehicles to transport cargo to and from wild destinations, not to mention large homes to store the goods associated with an outdoor lifestyle. The popularity of outdoor activities also generates demands for recreational resorts. Ski slopes and golf courses—and the residential developments that crop around them—can be a manifestation of a consumer impulse, which exacts a toll on the landscape. In these cases, enthusiasm for a place may harm rather than preserve the land. It is possible, after all, to love something to death.

If place has a particular importance in climate change and environmental education, it is not simply about listening to what "places are telling us." Rather, as Ruitenberg (2005, p. 218) argues in her "radical pedagogy of place," it is about coming to understand the significance of "the multiplicity of and conflicts between interpretations of a place, the traces of meanings carried by the place in the past, the openness to future interpretation and meaning-construction." Likewise, it is not principally about connecting with or fostering strong attachments to particular places over others, but learning how to live, work and exist in specific places in ways that don't exacerbate local or global climate and environmental crises, but contribute toward resolving these crises instead.

A second key issue is that the institution of formal schooling and broader structures and discourses of capitalist economy and society exert strong disciplining forces, and can easily colonize and redirect educational initiatives that may initially seek to explore and embrace place in pursuit of alternative, transformational social agendas (Gruenewald, 2004). If places can take on any number of different meanings, capitalism and formal education often play a dominant role in shaping what kinds of meaning become most powerful and prominent in any given place. As Robert Stevenson (2007, p. 148) warns, there is often a conflict between the transformational, solidaristic goals of much environmental education, and the concerns with control of individual behavior and development, and "competitive grading and ranking," that have long characterized the core grammar of formal schooling. When educators seek to bring in an engagement with place as a way to transform the unsustainable placelessness of traditional schooling, all too often, it is traditional schooling that ends up transforming these educational place engagements instead. We regularly see forest schools, school gardens and outdoor education promoted and justified for their ability to raise individual test scores and improve student behavior, as these are goals that are prioritized in most formal education systems—worthy outcomes, perhaps, but not centrally about addressing climate or environmental crises (Chawla et al., 2014; McCree, 2019; Williams & Dixon, 2013). "When environmental education is promoted as a vehicle (*using* the environment) for closing the achievement gap, rather than as a pathway to ecological literacy or to a more sustainable society," argue Gruenewald and Manteaw (2007, pp. 176–177), it risks becoming "muted, distorted and absorbed by the culture of schooling."

Hursh, Henderson and Greenwood (2015, p. 299) note how these pressures and contradictions have increased in the current neoliberal era, as much "environmental education is shaped by the political, cultural, and economic logic of neoliberalism," that favors individualism, consumerism and marketization. As Pudup (2008) writes of the well-known Edible Schoolyard Project in Berkeley, California, the focus of much school gardening in the contemporary period is to pursue social change through changing the attitudes and behaviors of students *as individual consumers*. Garrett Broad (2016, p. 175) calls this the "magic carrot" approach, that he argues is "fundamentally ill-equipped to engage with the broader economic, environmental, and sociostructural barriers" that are at the root of climate, environment, food and health crises. When, in early 2020, the Edible Schoolyard pledged to "help Americans eat their way out of climate change," its approach remained the same as it had been previously, when its focus was on issues of student diet and health. In the words of Edible Schoolyard teacher Esther Cook, the Project seeks to use school gardening to get students to "think of themselves as consumers, and [understand] that the choices they make as consumers have repercussions out in their own lives and in the real world" (Bowen, 2020). Not only does this focus on individual consumer behavior individualize responsibility for the climate crisis, it fails to recognize the root causes of the crisis or provide students with the collective, political means to tackle these root causes (Maniates, 2001). Individual consumers directly consume only a minority of the fossil fuels that contribute to the climate crisis (fossil fuels are also consumed through industrial production processes); and even when "individuals consume fossil fuels directly … they do so in the context of social and economic systems over which they may have little control" (Pirani, 2018, p. 15). The broader concern, then, is that rather than doing something radically different through an intensive, experiential engagement with concrete places, educators can be pressured into adopting more of the same kinds of ideas and practices that are currently dominant in formal education and neoliberal capitalist society—and more of the same is precisely what is not needed to address the climate crisis. Place matters for climate change education, but only if and when it is used effectively to analyse, understand, contest and transform the dominant patterns and practices of formal schooling and (neoliberal) capitalist economy that are currently working to create and exacerbate climate crisis.

Connecting with Nature?

Beyond these general concerns, there are specific questions about how we need to think about the importance of place in climate change and environmental education. One has to with the idea of using place to (re) make a "connection with nature" for students, that has become a principal goal in much environmental education. The central argument of the "connecting with nature" framework is that people, and especially children, have become increasingly disconnected from nature in the modern world, and this is creating all kinds of individual, social and environmental harms: poor physical and mental health and wellbeing, lower cognitive performance, increased social fragmentation and alienation, and a loss of pro environmental attitudes and behaviors (Charles et al., 2018; Chawla, 2015, 2020). One of the best known articulations of this argument is Richard Louv's suggestion that many children today suffer from a "nature deficit disorder." As Louv (2005, pp. 32, 92) writes:

> at the very moment that the bond is breaking between the young and the natural world, a growing body of research links our mental, physical, and

> spiritual health directly to our association with nature—in positive ways. … Nature-deficit disorder describes the human costs of our alienation from nature, among them: diminished uses of the senses, attention difficulties, and higher rates of physical and emotional illnesses.

Concern with the separation of children from nature long precedes Louv, and goes back to the origins of modern, Western conceptions of childhood (Kraftl et al., 2019): the rise of the Scouting movement in the early twentieth century, for example, was concerned with redressing the harm believed to be done to young boys at the time by rising urbanization and a growing separation from nature (MacDonald, 1993).

The remedy to this problem is to enable people (or children) to spend more time "in nature," on the assumption this will increase their sense of "nature connectedness," which in turn, will led to direct benefits, including a greater willingness and interest in taking action to address the climate crisis. Concern over the increasing "degree of *dis*connection [from nature] has led to the development of whole industries—academic, practitioner and media—around the steps that might be taken to (re)connect children and young people with nature" (Kraftl et al., 2019, p. 301; see also Taylor, 2013). National parks and forests, and global environmental, conservation and wildlife organizations now often run connecting with nature programs. In the United States, Richard Louv co-founded the Children & Nature Network in 2006, an umbrella organization that seeks to create "a world in which all children play, learn and grow in nature in their everyday lives" (Children & Nature Network, 2020); and in 2016, the International Union for Conservation of Nature launched the #NatureForAll campaign, as a "global movement to inspire love of nature," arguing that "personal experiences and connections with nature provide powerful benefits for individual and societal health, well-being, and resilience," as well as enabling "the achievement of … sustainable development goals and targets" (IUCN, 2020). A growing set of "instruments" have been constructed by researchers to measure how connected to nature an individual is, including the Connection to Nature Scale, the Nature Relatedness Scale, the Inclusion of Nature with Self, Environmental Identity Scale, and Emotional Affinity to Nature Scale (Hughes et al., 2018, p. 12).

Connecting with nature focuses on many issues other than climate change, but with the growing concern with climate crisis, it has been tied to this issue as well. What is the link between a connection with nature and action to address the climate crisis? The claim is that "caring for nature is a result of connecting with it," and that there is a "relationship between experiences in nature and the development of the sense of connectedness that tends to lead people to take action to care for nature" (Charles et al., 2018, pp. 7, 14). As Russell (1999, p. 123) writes, the proposed chain of causation is "NATURE EXPERIENCE -> CARING -> COMMITMENT -> ACTION." Selby and Kagawa (2018, p. 313) argue that any effective program of transformative climate change education will foster "enrichment through nature immersion, and emotional engagement with nature," so that each and every learner will:

> become motivated to feel affinity and intimacy with nature, to come to care for nature and to be ready to proactively defend nature against the debilitation and devastation that it so clearly faces from climate change and interrelated threats. … By knowing nature intimately, in different moods and seasons, in its "lived reality," in its details and particulars, the learner can come to love a natural place and fight for its preservation.

Selby and Kagawa emphasize the importance of "appreciating the intrinsic value of nature," and learning to feel its "beauty, attunement, awe, ecstasy, enchantment, reverence, rhapsody, joy and wonder" (p. 313).

While there are no doubt many benefits that can come from the proliferation of connecting with nature agendas and programs, there are also problems with this framework that demand attention. Critics question how nature is conceptualized in this framework, arguing that romanticized, western, white, middle class cultural understandings and habits of interacting with nature as wilderness, ideally untouched and uninhabited by humans, are assumed (Dickinson, 2013; Malone, 2016; Nxumalo & Cedillo, 2017). Certainly, if we adopt Soper's (1995, pp. 132–133) definition of nature as "those material structures and processes that are independent of human activity (in the sense that they are not a humanly created product), and whose forces and causal powers are the necessary conditions of every human practice, and determine the possible forms it can take," it is highly peculiar and problematic to claim that individuals have become disconnected from nature and are not always already existing within nature (as quoted in Malm, 2018). The connecting to nature approach articulates an anxiety over and disregard of urban areas, often failing to recognize fully the ways in which those "who live in urban environments are connected in all sorts of ways to diverse natures," that include quite unromantic elements such as "sewage systems, dense populations on scarce land, floods, high-rises, decay, contrived parks, shops, lights, pigeons, rats, waste disposal systems, street dogs and dog parks" (Hadfield-Hill & Zara, 2019, p. 68; Rautio et al., 2017, p. 1386). There is a perversity when environmental educators feel compelled to transport students across long distances in order to help them "find" and "connect" with "nature" (Harrison, 2010).

Critics also question whether time spent in wilderness areas and green spaces, and feelings of connection with nature, are always beneficial for addressing the climate and other environmental crises. As Chang (2017) suggests above, enthusiasm for outdoor activities in wilderness areas can be environmentally destructive. Kari Norgaard's (2011) *Living in Denial*, an ethnographic study of climate change attitudes in a rural community in Norway, describes how nature experiences and connectedness not only did not generally lead to increased action among community members to address the climate crisis, but provided a comforting excuse not to participate in any such climate action:

> Emphasizing a connection to nature … serves as a tool of innocence. If one of the things that made it hard for community members to think about climate change … was awareness of their own contribution to the problem, … then to reaffirm, through rituals, that [the community was] close to nature is to imply that despite their rising materialism, petroleum development, and wealth, they too are pure, naturally good, even 'natural' environmentalists. In the context of romantic notions of nature as pure, an association with nature provides a reassurance of original innocence. Relationships to nature are a way to legitimate Norwegian "goodness" in the face of antienvironmental behavior. (p. 149)

Claudio Rosa and Silvia Collado's (2019) research review emphasizes that the relationship between nature experiences and pro-environmental attitudes and actions is complex, and can depend on the types of experiences individuals have in natural areas, the activities they undertake while there, the different types of natural environments that are experienced, as well as the sociodemographic characteristics of those doing the nature experiencing.

Finally, as Dickinson (2013, p. 315) suggests, the idea that fostering a sense of individual connection with nature is a primary factor for addressing environmental and climate crisis is in many ways a "misdiagnosis" that can "obscure and mistreat the problem." Like many dominant approaches to environmental and climate change education, the connecting with nature framework risks individualizing responsibility for the climate crisis, while "displacing attention from the overarching political economy of ecological degradation" (Fletcher, 2017, p. 227). More than this, the framework risks falsely blaming those who have the least ability to get to "nature" as being the most problematic actors in the climate crisis. For spending time outdoors in "designated nature areas," as proposed by many connecting with nature programs, typically "requires transportation, time, money, consumption" and "physical and social access" that are not available to many "financially constrained, and geographically (and often racially) segregated families" (Dickinson, 2013, p. 322). The root cause of the climate crisis, as many argue, is not principally about whether individuals spend or don't spend time in wilderness areas or green spaces, nor whether they feel a particular connection with nature as wilderness. Rather, it is about an economic and cultural system that approaches nature instrumentally as something that can be endlessly exploited, dominated, controlled, depleted, polluted and destroyed without consequence—what Naomi Klein (2014) and others call a system based on "extractivism." Individuals' sense of disconnection from nature is a symptom of this economic and cultural system; and it is learning why we need to and how we can transform this system, not just experiencing and appreciating time spent in green spaces and wilderness areas, that is pivotal.

This raises one of the most problematic misdirections of the connecting with nature approach to the climate crisis: the suggestion that addressing this crisis is primarily about saving, defending or protecting nature. As Andreas Malm (2018) observes, nature is:

> the part of the inhabited world that humans encounter *but have not constructed*, created, built or conjured up in their imagination. … It preceded us, surrounds us and will succeed us; it was, is and will be spontaneously generated without us; it may be under all sorts of influence, but that does not put an end to it, any more than a continent ceases to be because it has skyscrapers standing on it.

Addressing the climate crisis is fundamentally about protecting different communities in society, by preserving those biophysical processes and cycles in nature that enable our collective and continued wellbeing and existence, and that of other species that cohabit the planet with us. It is about saving cities threatened by rising sea levels as much as it is about saving rainforests, aiding refugees uprooted and displaced by drought as much as polar bears suffering a loss of sea ice. It is about understanding and transforming the relationship *between* humanity and the natural world, from one based on endless extraction and domination, to one that enacts reciprocity and recognizes interdependence. In this, any framing of climate change education and action as being about "defending nature" can be limiting and harmful, especially when nature is conceptualized as wilderness removed from human population. It may be that for some the route to climate crisis action will happen through intense learning engagements with non-urban places of relative and remote wilderness. But for climate change education and action to develop at scale, it needs to happen through close learning engagements with the urban and suburban as well. All kinds of places matter for an effective climate change education, not just wild, green and rural ones.

Land Education and the Damages of Colonialism

One of the most devastating critiques of the connecting with nature framework is that the green spaces and wilderness areas to which (often white, middle class) students are taken to "connect with nature," have in many areas of the world been created through the systematic, racist and violent removal and exclusion of Indigenous and other residents who used to live on this land. Stanley Park in Vancouver, for example, one of North America's largest urban parks, where city residents are now encouraged to visit and "connect with nature," was created through the forced eviction and destruction of homes in the late 1800s of coastal Indigenous communities, who had lived in the area for hundreds of years previously (Barman, 2005; Eragoda, 2020). Likewise, many of the national forests and parks across the United States—now enthusiastically promoted by some as having "unparalleled potential ... to inspire civic engagement in climate change through place-based communication" (Schweizer et al., 2013, p. 42)—were created through the forced dispossession of their Indigenous inhabitants, in the name of fashioning "living Edens containing beautiful scenery, rivers, animals, flowering trees, and carpets of wildflowers" that could become "'virgin' places of rebirth in which people could be spiritually renewed" (Merchant, 2003, p. 382). Nor is this phenomenon found solely in settler colonialist states. The sparsely populated Scottish Highlands, where today outdoor educators take students on excursions to develop their "personal relationship with nature" (Harrison, 2010, p. 3), were produced by the forced eviction of local inhabitants during the Highland Clearances of the eighteenth century as part of the broader enclosure movement that swept across Europe at the time (Standing, 2019). Around the world, the fortress conservation model of creating protected spaces for pristine nature has led and continues to lead to the exclusion and dispossession of millions of Indigenous peoples from their ancestral lands. It remains, as Dominguez and Luoma (2020, p. 6) write, a "lasting legacy of colonialism in Africa, Asia and elsewhere." Critics argue that not only do connecting with nature and other environmental education approaches often erase and ignore these ongoing histories of colonization and expulsion, but in themselves, repeat and continue classic acts of settler colonialism, by rendering invisible Indigenous presence and claims, while enacting their own assertions of a renewed connection to and identity with the land—a process that Tuck et al. (2014, p 15) criticize as "settler emplacement."

From the perspective of developing transformative climate change education, there are three key issues raised by this critique. One is that when bringing place to the centre of learning, we need to avoid "apolitical and ahistorical approaches to place," by recognizing that most places have long, continuing histories of contestation and conflict over meaning, purpose, power, value and ownership, and recalling who and what used to live in and occupy places, but have since been made into "absent presences," marginalized, excluded, destroyed or made extinct (Nxumalo, 2019, pp. 18, 39). Learning in and about places, as Fikile Nxumalo (2019, p. 132) argues, "necessitates political choices, such as which stories, encounters, and histories to make visible, what knowledge counts in the making of a place, and which pastpresent inhabitants of place matter." The political process of excavating or refiguring absent presences addresses not just Indigenous and local human communities, but nonhuman species of animals and plants, as well as the land itself. Given that much of the destruction of the global environment and warming of the climate has occurred since the 1950s, part of any transformative climate change education needs to develop a strong historical memory and analysis beyond our own lifetimes of what has changed and been lost in particular places around the world and why (Angus, 2016). What we might have grown up with and thought of as "normal" in our own local places might actually be the outcome of a long history of destruction, violence, injustice and harm—a phenomenon known as "shifting baseline syndrome," in which environmental degradation and social injustice becomes taken for granted (Roberts, 2020; Soga & Gaston, 2018). Rather than developing strong attachments and commitments to defending and preserving places as these exist today, we often need instead to understand what was once in these places before, and what could and should be there again in the future (Calderon, 2014; Ruitenberg, 2005).

Second, there are close historical links between the spread of imperialism and colonialism, dispossession and expulsion of Indigenous communities, and acceleration of global environmental destruction and climate crisis. These are not separate concerns of injustice, incidentally linked, but are integrally connected to the same core set of structural and cultural problems and phenomena. Hannah Holleman (2018, p. 149), for example, traces a direct line from the global crisis of soil erosion in the early twentieth century to the worsening global environmental and climate crises of today, arguing that all of these crises are driven by an imperialist and colonialist model that combines ecological degradation with the "domination of peoples and attempted decimation of entire cultures."

This model, Holleman argues, is based on an ideology of "white supremacy" and "culture of conquest" that has provided a sense of legitimacy, purpose and righteousness to expansionist projects of seizure, domination and exploitation of land, nature and Indigenous peoples worldwide (p. 57). "The ideological legacy of the 'heyday of imperialism and colonialism,'" writes Holleman, "helps normalize global inequality today and facilitates the world's most powerful corporations' extraction of resources, labor, and financial wealth from the so-called least-developed regions" (p. 10). "Settler colonialism … not only violently interrupted Indigenous life," note Eve Tuck, Marcia McKenzie and Kate McCoy (2014, p. 19), it also "resulted in 'quick and brutal' environmental degradation" (see also Katz-Rosene & Paterson, 2019; Voskoboynik, 2018; Yusoff, 2018). Naomi Klein (2014, pp. 169–170) uses the concept of global "sacrifice zones" to describe this process:

> Sacrifice zones [are] places that, to their extractors, somehow don't count and therefore can be poisoned, drained, or otherwise destroyed, for the supposed greater good of economic progress. This toxic idea has always been intimately tied to imperialism, with disposable peripheries being harnessed to feed a glittering center, and it is bound up too with notions of racial superiority, because in order to have sacrifice zones, you need to have people and cultures who count so little that they are considered deserving of sacrifice. Extractivism ran rampant under colonialism because relating to the world as a frontier of conquest—rather than as a home—fosters this particular brand of irresponsibility. The colonial mind nurtures the belief that there is always somewhere else to go to and exploit once the current site of extraction has been exhausted.

As the history of wilderness areas and green spaces forewarns, contemporary efforts to "save nature" and address the climate crisis can also be shaped by this colonialist legacy of land seizure and Indigenous dispossession. The IPCC, for example, has promoted the climate saving technology of bio-energy with carbon capture and storage (or BECCS), which involves creating "massive tree plantations around the world" that can be used to "suck CO2 out of the atmosphere as they grow," and subsequently as biofuels can be harvested and burned for energy, with their carbon emissions captured and stored underground "where it can never escape" (Hickel, 2020, p. 129). However, as Jason Hickel (2020, p. 132) points out, for BECCS to operate at the scale called for by IPCC modelling, there would need to be "biofuel plantations covering an area two to three times the size of India, gobbling up about two-thirds of the planet's arable land." This, Hickel warns, could trigger "a kind of climate colonialism," as powerful nations seize land for biofuels around the world, displacing local and Indigenous communities who currently rely on this land for food production (p. 132; see also Reisdorf, 2020).

Finally, as we recognize that climate crisis and environmental destruction are linked to an economic and cultural system based on exploitation and domination of nature, some argue that Indigenous systems of knowledge can provide vital alternative models to learn from. Many Indigenous cultures embrace principles of what Nxumalo and Berg (2020, p. 50) call "radical relationality," understood as "an ethical orientation that requires reciprocity and obligation between current and future generations, lands, water, and animal and plant life;" many also have strong traditions of living sustainably within local places, based on deep, experiential knowledge of and cultural engagement with these places (see also Lowan-Trudeau, 2012; Pierotti, 2011; Shava, 2012). The concept of "Indigenous"—defined literally as "originating or occurring naturally in a particular place" (Stevenson, 2015)—is one that needs to be critically unpacked, for it groups together a range of radically different cultures and communities. There are also problematic traditions in western scholarship of treating Indigenous cultures as unchanging repositories or museums of information, rather than recognizing them to be, like all cultures, dynamic, changing, hybrid, contested and political; claiming that all Indigenous cultures and individuals have an essential and integral orientation towards sustainable ecological practice (the stereotype of the "ecological noble savage"); and constructing and exploiting Indigenous cultures instrumentally, for political, material and financial gain (Dove, 2006; Guenther et al., 2006; Hames, 2007; Rodrigue-Allouche, 2015). Nonetheless, Indigenous cultures represent a vast richness of knowledge and practice about "ways of being ecological in an astounding variety of places," and can offer "a wealth of clues about what real ecological intelligence looks like" (Hickel, 2020, p. 256). To refer to educational approaches that embrace these critical interventions, Tuck et al. (2014, p. 13) propose the term "land education," which puts "Indigenous critiques of settler colonialism," and "Indigenous epistemological and ontological accounts of land at the center," paying close attention to "constructions and storying of land and repatriation by Indigenous peoples, documenting and advancing Indigenous agency and land rights."

Uneven Geographies and Environmental Justice

Common critiques of "place-based education," then, are that it often neglects urban spaces in favor of rural and wilderness areas, embraces a problematic idea of "connecting to nature," is often apolitical and ahistorical, and ignores concerns of colonialist dispossession and exclusion of Indigenous and other marginalized communities. The embrace of place in environmental and place-based approaches to education also often takes a romanticized view of place as being inherently appealing, attractive, desirable and harmonious. But, as McInery et al. (2011, p. 10) point out, many "environments are far from idyllic places for children and their families," and though it may be "easy to feel a strong sense of attachment to an aesthetically pleasing landscape—a pristine stream, a beautiful valley or a leafy-green suburb," things might be different if one lives in an "unsafe, environmentally degraded place or one that is fractured by social, economic and racial divides." There is, however, another long tradition of environmental action and learning that also puts place at the center of its approach, to which these criticisms do not apply: this is the tradition of environmental justice, which provides an essential corrective to thinking about place in any effective, transformative climate change education (Agyeman et al., 2016).

Environmental justice, which is concerned with social mobilizing to redress unjust and unequal imposition of environmental harms on low income and non-white communities around the world, takes a markedly different approach to conceptualizing the meaning of "environment" to that found in many mainstream environmental organizations (Schlosberg

& Collins, 2014). Rather than focus on wilderness areas and green spaces, environment is defined in human terms as "where we live, where we work, where we play, and where we learn," and tied closely to ideas of "home and community" (Cole & Foster, 2001, p. 2016; Novotny, 1995). Environmental justice is concerned with the links between environmental issues and social justice demands for equal access to high quality housing, employment, education, health, transportation and civil rights protections; and with the systematic, structural and institutional inequalities of race, class, nation and gender that shape who usually bears the brunt of environmental harms and hazards (Mascarenhas, 2021; Sze, 2020). The places that attract the most attention from environmental justice academics and activists are often urban and suburban, marked by patterns of racism, sexism, colonialism, exploitation, marginalization and exclusion, and suffering from extensive environmental degradation. These are Klein's (2014) sacrifice zones, "poor places," "out-of-the-way places," and "places where residents [lack] political power, usually having to do with some combination of race, language, and class": places such as hazardous waste sites, landfills, incinerators, refineries, chemical plants, mining operations, logging areas, flood plains and highways (Bullard, 2020). The argument of environmental justice is that these places are just as important to focus environmental attention on as any unpopulated space of wild nature.

For climate change education, environmental justice offers several key lessons. Common approaches to the climate crisis consider the crisis as a global phenomenon, and adopt what Malm and Warlenius (2019, p. 33) call a "universalist conception of climate change": they treat climate change as something that is caused by the combined actions of all humans, and conversely, as "a general process, affecting everyone everywhere" (Goh, 2020, p. 563). We see this in claims about the rise of the Anthropocene, as a new geological age in which humans have become the dominant influence in shaping the global climate and environment (Davis et al., 2019). It is central to approaches to climate change education that focus on individual behavior change, on the assumption that "human behavior is responsible for many of our greatest environmental challenges," including climate crisis (Byerly et al., 2018). We see this idea, too, in statements by groups like Extinction Rebellion (2020) that all of humanity is threatened by climate change, with "the possibility of human extinction in the near future." Environmental justice insists on taking a situated approach to the climate crisis, recognizing that different groups in different places have differing levels of responsibility in contributing to the climate crisis, and that the impacts of the crisis are and will continue to be radically uneven for different communities around the world (Stapleton, 2020). Climate crisis may impact us all, but it does not and will not affect us all equally: our place in the world matters centrally in how we experience this crisis.

Differences in race, class, nationality and gender are central to any understanding of and response to the climate crisis. This is the insight of climate justice organizing, that has developed out of the environmental justice movement (Schlosberg & Collins, 2014). Not only are low income and nonwhite communities likely to have contributed less to causing the climate crisis, they are more likely to be impacted by its harms:

> People living in poverty are exposed to persistent, intersecting, and entrenched structural inequalities, making them particularly vulnerable to harm from the hazards unleashed by climate change. … People of color, Indigenous peoples, and women are examples of social groups that are disproportionately vulnerable to the long-term effects of increased air pollution, extreme heat, drought, food and water shortages, infectious disease, storms, and floods. They are more vulnerable to climate disruption due to discrimination, cultural expectations, and subordinate positions in social hierarchies. (Harlan et al., 2015, p. 16; see also Islam & Winkel, 2017)

These inequalities may be seen at a range of spatial scales, local, regional, national and global. Much political and media attention have focused on inequalities between nations—for example, the plight of low lying island nations that, though having little responsibility for creating the climate crisis, are threatened with existential obliteration through rising sea levels caused by the crisis (Baptiste & Rhiney, 2016; Shenk, 2012). But inequalities and injustices are also found locally, so that within a single city, low income and nonwhite neighborhoods are frequently more vulnerable to the hazards of extreme heat caused by climate change than wealthy and white neighborhoods, due to factors like geographical location, low quality housing stock and lack of tree cover and green space (Harlan et al., 2019).

Finally, when we consider the importance of place in climate change education, our perspective has to be comparative, comparing the places in which we live, work, play and learn with other places around the world. Such a comparative approach is central to the environmental justice approach, for it is the only way that we can recognize the existence of uneven geographies in the world, those structural patterns of inequality and injustice that differentiate places of privilege from places of neglect and marginalization, and lead to low income and non white communities bearing the brunt of environmental hazard and harm. When Laura Pulido (2017, p. 524) critiques the existence of an "environmental racism gap," she bases this on a comparative analysis between places that finds "compelling evidence that environmental disparities between white and nonwhite communities … have not diminished and that the situation may have worsened." If we fail to adopt this comparative approach and develop a keen understanding of how our places may differ to those of others, and why, there is a risk that learning and action to address the climate crisis, rather than successfully address the root causes of the crisis as a whole, will instead reproduce the patterns of inequality and injustice that have contributed to creating the crisis to begin with. Climate crisis learning and action is then at risk of perpetuating what some call "climate apartheid" (Dawson, 2017): a world consisting of racially and socioeconomically segregated geographies where, "on one side of town, there [are] ecological 'haves,' enjoying access to healthy, morally upstanding green products and services," while "on the other side of town, ecological 'have-nots' [languish] in the smoke, fumes, toxic chemicals, and illnesses of the old pollution-based economy" (Van Jones, quoted in Ross, 2011, p. 70; see also Agyeman & Zavestoski, 2020).

Local Places in a Global Economy

One of the longstanding concerns about foregrounding a concern with place in education is the fear this will lead to turning away from global engagements. David Orr (1992, p. 97) acknowledges that "critics might argue that the study of place would be inherently parochial and narrowing." Jan Nespor (2008, p. 479) criticizes place-based education for adopting "an idealized image of 'place' as a stable, bounded, self-sufficient communal realm" that undermines our ability to recognize the global and translocal cultural, economic and political structures and forces that inevitably shape all of our lives, wherever in the world we are situated. "Restricting learning … to the schoolyard and local neighbourhood appears somewhat myopic," McInery et al. (2011, p. 10) warn, as "students can also learn much about themselves and their world by studying other cultures, places and times." Likewise, there has often been a worry that focusing on place in education implies privileging small scale localism over other kinds of activist organizing at regional, national and global levels, which might be essential for success in addressing problems such as the climate crisis (McInery et al., 2011). To address problems of local environmental injustice that are presented by global environmental crises such as the climate crisis, Hilda Kurtz (2003, p. 891) observes, there may be "a disjuncture between the geographic scale(s) at which a problem is experienced, and the scale(s) at which it can politically be addressed" (as quoted in Nespor, 2008). Even if there are clear advantages to a focus on place in climate change education, perhaps there are trade-offs and liabilities as well?

None of these concerns are necessary if we adopt what Doreen Massey (1994, pp. 154, 156) calls a "global sense of place," that instead of viewing place as a self-contained and bounded community and locale, recognises it instead as a "particular, unique, point of … intersection" of "networks of social relationships and movements and communications" that extend regionally, nationally and globally. "All places," as Massey (2007) points out:

> have lines that run out from them: trade routes, investments, political and cultural influences, … power relations of all sorts that run around the globe and that link the fate of other places to what is done in [this place]. This is the … external geography … of a global sense of place. For each place this geography, this tentacular stretching of power relations, will be particular.

In this view of place, educational engagement with place is precisely a way of encountering the global, a process of giving "curricular and pedagogical attention to 'unhiding' the ways people in different locations are linked," as "the *other* places to which [we] are connected, and in relation to which [we] are constituted" are often "hidden from our view, segregated from our everyday concerns, by circuits of communication, representation, and education" (Nespor, 2008, pp. 487–488). Further, there is nothing that requires our engagements with place to adopt one single (local) scale over others. Part of the purpose of engaging with place in climate change education is to construct meaningful, experiential, comprehensible and actionable ways of engaging with the climate crisis. But all of us regularly identify and engage with a range of different (local, regional, national, global) scales throughout our lives—the places that we attach to and work with can be both tiny and enormous (Devine-Wright, 2013). Any focus on place in climate change education will likely benefit from being multi-scalar in nature, not dogmatically local.

In considering local places in relation to global processes, the bias is to conceptualise these local places as being shaped by the global. "The local" is construed as a "product" of the global, writes Doreen Massey, as "the global" is "somehow always out there, … somewhere else in its origins," and "global forces arrive from 'elsewhere' and wreak havoc on a previous local embeddedness" (Massey, 2006, p. 65; 2007, pp. 16, 21; see also Tannock, 2010). This tends to be true in analyses of the climate crisis as well, as we look at how the global crisis has varying impacts on different places around the world. But the inverse process, of how actions in one place can impact global processes, as well as other places around the world, is just as important to consider. "People everywhere may be engaged in behaviors that affect others far beyond one's immediate locality," writes David Chang (2017, pp. 726–727), and a question that needs to be addressed in our educational engagements with place is "*How does this place affect other places?*" (italics in original). More than this, certain places can come to have exaggerated impacts on global processes such as climate change, and on environmental degradation in a wide range of other places across the planet. Global cities such as London and New York are home to a critical mass of global corporate and financial command and control centres and networks, and as such, are places where decisions and actions made locally can have "effects … even in the remotest corner of the globe" (Massey, 2007, p. 7). Likewise, as Klein (2014) writes, new "extractive projects" that "are attempting to dig and drill, whether for open-pit mines, or gas fracking, or tar sands oil pipelines," have environmental consequences that extend worldwide. A vital part of transformative climate change education, then, is to learn to analyse and understand these relationships of power, control and influence that exist between places around the world, and in so doing, open up what are often quite local points of action that can be taken with global consequences. For Massey (2004) and others, this kind of engagement with place is about constructing geographies of responsibility and accountability, to make clear what is often obscured not just by parochial accounts of the local, but abstract discussions of the global as well.

References

Agyeman, J., & Zavestoski, S. (2020). From dumping to displacement: New frontiers for just sustainabilities. In M. Mascarenhas (Ed.), *Lessons in environmental justice: From civil rights to Black Lives Matter and Idle No More*. Sage.

Agyeman, J., et al. (2016). Trends and directions in environmental justice: From inequity to everyday life, community, and just sustainabilities. *Annual Review of Environment and Resources, 41*, 321–340.

Angus, I. (2016). *Facing the Anthropocene: Fossil capitalism and the crisis of the earth system*. Monthly Review Press.

Baptiste, A., & Rhiney, K. (2016). Climate justice and the Caribbean: An introduction. *Geoforum, 73*, 17–21.

Barman, J. (2005). *Stanley Park's secret.* Harbour Publishing.

Bowen, K. (2020). Alice Waters' Edible Schoolyard Project changed how Berkeley students eat: Its next goal: To fight climate change. *Berkeleyside*, 4 February. https://www.berkeleyside.com/2020/02/04/alice-waters-edible-schoolyard-project-changed-how-berkeley-students-eat-its-next-goal-to-fight-climate-change

Broad, G. (2016). *More than just food: Food justice and community change.* University of California Press.

Bullard, R. (2020). From civil rights to Black Lives Matter. In M. Mascarenhas (Ed.), *Lessons in environmental justice: From civil rights to Black Lives Matter and Idle No More.* Sage.

Burningham, K., & Thrush, D. (2001). *'Rainforests are a long way from here': The environmental concerns of disadvantaged groups*. Joseph Rowntree.
Byerly, H., et al. (2018). Nudging pro-environmental behavior: Evidence and opportunities. *Frontiers in Ecology and the Environment, 16*(3), 159–168.
Cairns, K. (2018). Beyond magic carrots: Garden pedagogies and the rhetoric of effects. *Harvard Educational Review, 88*(4), 516–537.
Calderon, D. (2014). Speaking back to Manifest Destinies: A land-education-based approach to critical curriculum inquiry. *Environmental Education Research, 20*(1), 24–36.
Chang, D. (2017). Diminishing footprints: Exploring the local and global challenges to place-based environmental education. *Environmental Education Research, 23*(5), 722–732.
Charles, C., et al. (2018). *Home to us all: How connecting with nature helps us care for ourselves and the earth*. Children & Nature Network.
Chawla, L. (2015). Benefits of nature contact for children. *Journal of Planning Literature, 30*(4), 433–452.
Chawla, L. (2020). Childhood nature connection and constructive hope: A review of research on connecting with nature and coping with environmental loss. *People & Nature*, 1–24. https://doi.org/10.1002/pan3.10128
Chawla, L., et al. (2014). Green schoolyards as havens from stress and resources for resilience in childhood and adolescence. *Health & Place, 28*, 1–13.
Children & Nature Network. (2020). Vision & Mission. https://www.childrenandnature.org/about/
Cole, L., & Foster, S. (2001). *From the ground up: Environmental racism and the rise of the environmental justice movement*. NYU Press.
Corbett, M. (2009). Rural schooling in mobile modernity: Returning to the places I've been. *Journal of Research in Rural Education, 24*(7), 1–13.
Cresswell, T. (2015). *Place*. Wiley.
Davis, J., et al. (2019). Anthropocene, Capitalocene, … Plantationocene? A manifesto for ecological justice in an age of global crises. *Geography Compass, 13*, 1–15.
Dawson, A. (2017). *Extreme cities*. Verso.
Devine-Wright, P. (2013). Think global, act local? The relevance of place attachments and place identities in a climate changed world. *Global Environmental Change, 23*, 61–69.
Dickinson, E. (2013). The misdiagnosis: Rethinking 'nature-deficit-disorder'. *Environmental Communication, 7*(3), 315–335.

Dominguez, L., & Luoma, C. (2020). Decolonising conservation policy: How colonial land and conservation ideologies persist and perpetuate indigenous injustices at the expense of the environment. *Land, 65*(9), 1–22.

Dove, M. (2006). Indigenous people and environmental politics. *Annual Review of Anthropology, 35*, 191–208.

Eragoda, M. (2020). Tent cities and the violent origins of Vancouver's parks. *The Tyee*, 28 July. https://thetyee.ca/Opinion/2020/07/28/Violent-Origins-Vancouver-Parks/

Extinction Rebellion. (2020). Still heading for extinction and what to do about it. https://extinctionrebellion.uk/event/still-heading-for-extinction-and-what-to-do-about-it-3/

BC Farms & Food. (2019). School gardens: Preparing kids for climate change. 23 October. https://bcfarmsandfood.com/school-gardens-preparing-kids-climate-change/

Fletcher, R. (2017). Connection with nature is an oxymoron: A political ecology of 'nature-deficit-disorder'. *Journal of Environmental Education, 48*(4), 226–233.

Goh, K. (2020). Urbanising climate justice: Constructing scales and politicising difference. *Cambridge Journal of Regions, Economy and Society, 13*(3), 559–574.

Green, A. (1990). *Education and state formation*. Palgrave Macmillan.

Greenwood, D. (2009). Place: The nexus of geography and culture. In M. McKenzie et al. (Eds.), *Fields of green: Restorying culture, environment, and education* (pp. 271–282). Hampton Press.

Gruenewald, D. (2003). Foundations of place: A multidisciplinary framework for place-conscious education. *American Educational Research Journal, 40*(3), 619–654.

Gruenewald, D. (2004). A Foucauldian analysis of environmental education: Toward the socioecological challenge of the Earth Charter. *Curriculum Inquiry, 34*(1), 71–107.

Gruenewald, D., & Manteaw, B. (2007). Oil and water still: How No Child Left Behind limits and distorts environmental education in US schools. *Environmental Education Research, 13*(2), 171 188.

Guenther, M., et al. (2006). The concept of indigeneity. *Social Anthropology, 14*(1), 17–32.

Hadfield-Hill, S., & Zara, C. (2019). Complicating childhood-nature relations: Negotiated, spiritual and destructive encounters. *Geoforum, 98*, 66–74.

Hames, R. (2007). The ecologically noble savage debate. *Annual Review of Anthropology, 36*, 177–190.

Harlan, S., et al. (2015). Climate justice and inequality. In R. Dunlap & R. Brulle (Eds.), *Climate change and society: Sociological perspectives*. Oxford University Press.

Harlan, S., et al. (2019). Pathways to climate justice in a desert metropolis. In L. Mason & J. Rigg (Eds.), *People and climate change: Vulnerability, adaptation, and social justice* (pp. 23–50). Oxford University Press.

Harrison, S. (2010). 'Why are we here?' Taking 'place' into account in UK outdoor environmental education. *Journal of Adventure Education & Outdoor Learning, 10*(1), 3–18.

Hayhoe, K. (2017). Yeah, the weather has been weird. *Foreign Policy*, 31 May. https://foreignpolicy.com/2017/05/31/everyone-believes-in-global-warming-they-just-dont-realize-it/

Hickel, J. (2020). *Less is more: How degrowth will save the world.* William Heinemann.

Holleman, H. (2018). *Dust bowls of empire: Imperialism, environmental politics, and the injustice of 'green' capitalism*. Yale University Press.

Hughes, J., Richardson, M., & Lumber, R. (2018). Evaluating connection to nature and the relationship with conservation behaviour in children. *Journal for Nature Conservation, 45*, 11–19.

Hulme, M. (2008). Geographical work at the boundaries of climate change. *Transactions of the Institute of British Geographers, 33*, 5–11.

Hursh, D., Henderson, J., & Greenwood, D. (2015). Environmental education in a neoliberal climate. *Environmental Education Research, 21*(3), 299–318.

IPCC—Intergovernmental Panel on Climate Change. (2018). *Global warming of 1.5°C.* IPCC.

Islam, S., & Winkel, J. (2017). *Climate change and social inequality*. DESA Working Paper No. 152. New York: UN Department of Economic and Social Affairs.

IUCN—International Union for Conservation of Nature. (2020). #NatureForAll. https://www.iucn.org/commissions/commission-education-and-communication/our-work/natureforall

Katz-Rosene, R., & Paterson, M. (2019). Imperialism and environment. In I. Ness & Z. Cope (Eds.), *The Palgrave encyclopedia of imperialism and anti-imperialism* (pp. 1–11). Palgrave Macmillan.

Khadka, A., et al. (2021). Unpacking the power of place-based education in climate change communication. *Applied Environmental Education & Communication, 20*(1), 77–91.

Klein, N. (2014). *This changes everything*. Penguin.

Kraftl, P., et al. (2019). (Re)thinking (re)connection: Young people, 'natures' and the water-energy-food nexus in Sao Paulo State, Brazil. *Transactions of the Institute of British Geographers, 44*, 299–314.
Kudryavtsev, A., Stedman, R., & Krasny, M. (2012). Sense of place in environmental education. *Environmental Education Research, 18*(2), 229–250.
Kurtz, H. (2003). Scale frames and counter-scale frames: Constructing the problem of environmental injustice. *Political Geography, 22*, 887–916.
Louv, R. (2005). *Last child in the woods: Saving our children from nature-deficit-disorder*. Algonquin Books.
Lowan-Trudeau, G. (2012). Indigenous environmental education research in North America. In R. Stevenson et al. (Eds.), *International handbook of research on environmental education* (pp. 404–408). Routledge.
MacDonald, R. (1993). *Sons of the Empire: The frontier and the Boy Scout Movement, 1890–1918*. University of Toronto Press.
Mah, K. (2010). Young gardeners: On gardens as spaces of experiential pedagogy. *Public, 41*, 99–107.
Malm, A. (2018). *The progress of this storm: Nature and society in a warming world*. Verso.
Malm, A., & Warlenius, R. (2019). The grand theft of the atmosphere: Sketches for a theory of climate injustice in the Anthropocene. In K. Bhavnani et al. (Eds.), *Climate futures: Re-imagining global climate justice*. Zed Books.
Malone, K. (2016). Reconsidering children's encounters with nature and place using posthumanism. *Australian Journal of Environmental Education, 32*(1), 1–15.
Malone, K. (2017). Place. In M. Peters (Ed.), *Encyclopedia of educational philosophy and theory* (pp. 1883–1888). Springer.
Maniates, M. (2001). Individualization: Plant a tree, buy a bike, save the world? *Global Environmental Politics, 1*(3), 31–52.
Mascarenhas, M. (Ed.). (2021). *Lessons in environmental justice: From civil rights to Black Lives Matter and Idle No More*. Sage.
Massey, D. (1994). *Space, place and gender*. Polity Press.
Massey, D. (2004). Geographies of responsibility. *Geografiska Annaler, 86B*(1), 5–18.
Massey, D. (2006). London inside-out. *Soundings, 32*, 62–71.
Massey, D. (2007). *World city*. Polity Press.
McCree, M. (2019). When forest school isn't forest school. In M. Sackville-Ford & H. Davenport (Eds.), *Critical issues in forest schools* (pp. 3–20). Sage.

McInery, P., Smyth, J., & Down, B. (2011). 'Coming to a place near you?' The politics and possibilities of a critical pedagogy of place-based education. *Asia-Pacific Journal of Teacher Education, 39*(1), 3–16.
Merchant, C. (2003). Shades of darkness: Race and environmental history. *Environmental History, 8*, 380–394.
Monroe, M., et al. (2019). Identifying effective climate change education strategies: A systematic review of the research. *Environmental Education Research, 25*(6), 791–812.
Nakagawa, Y., & Payne, P. (2015). Critical place as a fluid margin in post-critical environmental education. *Environmental Education Research, 21*(2), 149–172.
Nespor, J. (2008). Education and place: A review essay. *Educational Theory, 58*(4), 475–489.
Noddings, N. (2005). Place-based education to preserve the earth and its people. In N. Noddings (Ed.), *Educating citizens for global awareness* (pp. 57–69). Teachers College Press.
Norgaard, K. (2011). *Living in denial: Climate change, emotions, and everyday life*. MIT Press.
Novotny, P. (1995). Where we live, work and play: Refraining the cultural landscape of environmentalism in the environmental justice movement. *New Political Science, 16*(1), 61–79.
Nxumalo, F. (2019). *Decolonizing place in early childhood education*. Routledge.
Nxumalo, F., & Berg, L. (2020). Conversations on climate change pedagogies in a Central Texas kindergarten classroom. In J. Henderson & A. Drewes (Eds.), *Teaching climate change in the United States*. Routledge.
Nxumalo, F., & Cedillo, S. (2017). Decolonizing place in early childhood studies: Thinking with Indigenous onto-epistemologies and Black feminist geographies. *Global Studies of Childhood, 7*(2), 99–112.
Orr, D. (1992). *Ecological literacy*. State University of New York Press.
Pierotti, R. (2011). *Indigenous knowledge, ecology, and evolutionary biology*. Routledge.
Pirani, S. (2018). *Burning up*. Pluto.
Pudup, M. (2008). It takes a garden: Cultivating citizen-subjects in organized garden projects. *Geoforum, 39*, 1228–1240.
Pulido, L. (2017). Geographies of race and ethnicity II: Environmental racism, racial capitalism and state-sanctioned violence. *Progress in Human Geography, 41*(4), 524–533.
Rautio, P., et al. (2017). Reconfiguring urban environmental education with 'shitgull' and a 'shop'. *Environmental Education Research, 23*(10), 1379–1390.

Reisdorf, P. (2020). Biofuels: Another story of neo-colonialism. *Discover Society*, 1 July. https://discoversociety.org/2020/07/01/viewpoint-biofuels-another-story-of-neo-colonialism/

RHS—Royal Horticultural Society. (2019). Use school gardens to fight climate change, says RHS. 6 September. https://www.rhs.org.uk/press/releases/Use-school-gardens-to-fight-climate-change-says-RH

Roberts, J. (2012). *Beyond learning by doing: Theoretical currents in experiential education*. Routledge.

Roberts, D. (2020). The scariest thing about global warming (and Covid-19). *Vox*, 7 July. https://www.vox.com/energy-and-environment/2020/7/7/21311027/covid-19-climate-change-global-warming-shifting-baselines

Rodrigue-Allouche, S. (2015). *Conservation and indigenous peoples: The adoption of the ecological noble savage discourse and its political consequences*. MA thesis, Uppsala University.

Rosa, C., & Collado, S. (2019). Experiences in nature and environmental attitudes and behaviors: Setting the ground for future research. *Frontiers in Psychology, 10*, 1–9.

Ross, A. (2011). *Bird on fire: Lessons from the world's least sustainable city*. Oxford University Press.

Rudiak-Gould, P. (2014). *Place-based approaches to climate education*. Pacific Islands Climate Education Partnership.

Ruitenberg, C. (2005). Deconstructing the experience of the local: Toward a radical pedagogy of place. In K. Howe (Ed.), *Philosophy of Education 2005* (pp. 212–220). Philosophy of Education Society.

Russell, C. (1999). Problematizing nature experience in environmental education: The interrelationship of experience and story. *The Journal of Experimental Education, 22*(3), 123–137.

Russell, J. (2019). Food, farms, and climate resilience: Bringing it home. *HEART*. https://teachheart.org/2019/12/17/food-farms-and-climate-resilience-bringing-it-home/

Said, E. (2000). *Reflections on exile*. Harvard University Press.

Schlosberg, D., & Collins, L. (2014). From environmental to climate justice: Climate change and the discourse of environmental justice. *WIREs Climate Change, 5*, 359–374.

Schweizer, S., Davis, S., & Thompson, J. (2013). Changing the conversation about climate change: A theoretical framework for place-based climate change engagement. *Environmental Communication, 7*(1), 42–62.

Selby, D., & Kagawa, F. (2018). Teetering on the brink: Subversive and restorative learning in times of climate turmoil and disaster. *Journal of Transformative Education, 16*(4), 302–322.
Shava, S. (2012). The representation of indigenous knowledges. In R. Stevenson et al. (Eds.), *International handbook of research on environmental education* (pp. 384–393). Routledge.
Shenk, J. (2012). *The island president*. Documentary film. London: Dogwoof.
Smith, G. (2020). Place-based education. In *Oxford research encyclopedia of education* (pp. 1–25). Oxford University Press.
Sobel, D. (2004). *Place-based education*. Orion.
Soga, M., & Gaston, K. (2018). Shifting baseline syndrome: Causes, consequences, and implications. *Frontiers in Ecology and the Environment, 16*(4), 222–230.
Soper, K. (1995). *What is nature? Culture, politics and the non-human*. Blackwell.
Standing, G. (2019). *Plunder of the commons*. Penguin.
Stapleton, S. (2020). Toward critical environmental education: A standpoint analysis of race in the American environmental context. *Environmental Education Research, 26*(2), 155–170.
Stevenson, R. (2007). Schooling and environmental education: Contradictions in purpose and practice. *Environmental Education Research, 13*(2), 139–153.
Stevenson, A. (Ed.). (2015). *Oxford dictionary of English*. Oxford University Press.
Stevenson, R., Mannion, G., & Evans, N. (2018). Pedagogies and place: An overview and analysis. In A. Cutter-Mackenzie et al. (Eds.), *Research handbook on childhoodnature* (pp. 1–21). Springer.
Stone, M. (2009). *Smart by nature: Schooling for sustainability*. Center for Ecoliteracy.
Sze, J. (2020). *Environmental justice in a moment of danger*. University of California Press.
Tannock, S. (2010). Learning to plunder: Global education, global inequality and the global city. *Policy Futures in Education, 8*(1), 82–98.
Taylor, A. (2013). *Reconfiguring the natures of childhood*. Routledge.
Trelstad, B. (1997). Little machines in their gardens: A history of school gardens in America, 1891 to 1920. *Landscape Journal, 16*(2), 161–173.
Tuck, E., McKenzie, M., & McKoy, K. (2014). Land education: Indigenous, post-colonial, and decolonizing perspectives on place and environmental education research. *Environmental Education Research, 20*(1), 1–23.

Voskoboynik, D. (2018). To fix the climate crisis, we must face up to our imperial past. *OpenDemocracy*. 8 October. https://www.opendemocracy.net/en/opendemocracyuk/to-fix-climate-crisis-we-must-acknowledge-our-imperial-past/

Wake, S., & Birdsall, S. (2016). Can school gardens deepen children's connection to nature? In T. Skelton, K. Nairn, & P. Kraftl (Eds.), *Space, place, and environment* (pp. 89–113). Springer.

Williams, D., & Dixon, P. (2013). Impact of garden-based learning on academic outcomes in schools: Synthesis of research between 1990 and 2010. *Review of Educational Research, 83*(2), 211–235.

Yusoff, K. (2018). *A billion black Anthropocenes or none*. University of Minnesota Press.

6

Children & Youth Can Change the World!

In 2011, the seventeenth annual United Nations Climate Change Conference was held in Durban, South Africa. The last speaker was a twenty-one year old youth delegate named Anjali Appadurai, who took the stage to chastize the climate negotiators for their endemic failures to take significant actions that would effectively address the climate crisis. "I speak for more than half the world's population," Appadurai said:

> We are the silent majority. You've given us a seat in this hall, but our interests are not on the table. … You've been negotiating all my life. In that time, you've failed to meet pledges, you've missed targets, and you've broken promises. … The most stark betrayal of your generation's responsibility to ours is that you call this 'ambition.' Where is the courage in these rooms?

A video of Appadurai's speech was posted online and went viral, watched tens of thousands of times. A review of youth climate activism would later argue that Appadurai's speech "was a defining moment for the youth climate justice movement" that had been "systematically silenced by those in power," since it "constituted a collective scream of resistance to the exclusivities and biases of the [global climate negotiations] process as

S. Tannock, *Educating for Radical Social Transformation in the Climate Crisis*, Palgrave Studies in Education and the Environment,
https://doi.org/10.1007/978-3-030-83000-7_6

well as its ineffectiveness in bringing about meaningful change" (Foran et al., 2017, p. 363). At the Durban conference itself, the session chairperson admitted that perhaps youth delegates such as Appadurai should have been given the floor at the beginning of the conference, rather than pushed to the very end (Democracy Now!, 2011; Goodman, 2011).

Almost two decades before Anjali Appadurai's Durban speech, it had been a twelve year old named Severn Cullis-Suzuki who spoke at the 1992 United Nations Earth Summit in Rio de Janeiro, earning a standing ovation from the audience. Like Appadurai, Cullis-Suzuki stated that she had come to the conference to represent the interests of the world's children and youth—"I am here to speak for all generations to come"—and like Appadurai, she was critical of the adult officials who were seated before her:

> In my anger, I am not blind and in my fear I am not afraid of telling the world how I feel. … At school, even in kindergarten, you teach us how to behave in the world. You teach us not to fight with others, to work things out, to respect others and to clean up our mess, not to hurt other creatures, to share, not be greedy. Then, why do you go out and do the things you tell us not to do? Do not forget why you are attending these conferences, who you are doing this for. We are your own children. You are deciding what kind of a world we are growing up in. Parents should be able to comfort their children by saying, 'Everything is going to be all right, it's not the end of the world, and we are doing the best we can.' But I don't think you can say that to us anymore. Are we even on your list of priorities? My dad always says, 'You are what you do, not what you say.' Well, what you do makes me cry at night. You grown-ups say you love us. But I challenge you, please, make your actions reflect your words.

After the speech, Cullis-Suzuki became something of an environmentalist child celebrity, dubbed "The Girl Who Silenced the World for Five Minutes," praised by world leaders, and interviewed by media outlets across the globe (Smith, 1993; Stainsby, 1993). Cullis-Suzuki published a book, *Tell the World: A Young Environmentalist Speaks Out* (1993); and was made the focus of a film, *Severn: The Voice of Our Children* (Jaud, 2010), and children's book, *Severn and the Day She Silenced the World* (Wilson, 2014).

More recently, in 2018, it was Greta Thunberg, then fifteen years old, whose speech at the twenty-fourth annual United Nations Climate Change Conference in Katowice, Poland went viral on social media. Like Appadurai and Cullis-Suzuki, Thunberg spoke in the name of all children and young people and was critical of the climate negotiators:

> You only talk about moving forward with the same bad ideas that got us into this mess. Even when the only sensible thing to do is to pull the emergency brake. You are not mature enough to tell it like it is. Even that burden you leave to us children. … You say that you love your children above everything else. And yet you are stealing their future.

Unlike Appadurai and Cullis-Suzuki, however, Thunberg did not end her speech with a plea for climate negotiators to take action. "We have not come here to beg world leaders to care," Thunberg said, since "you have ignored us in the past and you will ignore us again." Instead, she warned the negotiators that "we've come here to let you know that change is coming whether you like it or not," as "the real power belongs to the people" (Thunberg, 2019). Thunberg had come to the conference having already attained global attention for launching her school strike for climate campaign a few months earlier in Stockholm, a campaign that over the next two years would expand into the global Fridays for Future movement, with 14 million children and adult allies participating in 220,000 school strikes in 212 countries (Fridays for Future, 2020). On the basis of this mobilization, Thunberg became one of the world's best known spokespersons on the climate crisis, invited to address national governments, publishing two books—*No One is Too Small to Make a Difference* (Thunberg, 2019) and *Our House is on Fire* (Ernman et al., 2020)—made into the subject of numerous other books and films—including *Greta's Story: The Schoolgirl Who Went on Strike to Save the Planet* (Camerini, 2019), *Greta and the Giants* (Tucker & Persico, 2019), *The Greta Thunberg Story: Being Different is a Superpower* (Part, 2019) and *Planet Greta: How Greta Thunberg Wants You to Help Her Save Our Planet* (Stead & Baker, 2020)—nominated for the Nobel Peace Prize and named *Time* magazine's Person of the Year in 2019 (Witt, 2020).

This brief history of a single strand of child and youth environmental and climate activism highlights a number of important points. First, child and youth involvement in climate change education and action is not a new phenomenon. Children and young people have shown their ability and interest in speaking and acting to address environmental and climate crises for as long as these crises have attracted public attention, and repeatedly demanded a public and political role regularly denied to them. Nor are the stories of Appadurai, Cullis-Suzuki or Thunberg isolated cases. What has become clear, particularly since the rise to global prominence of Thunberg, is that there are thousands of children and youth around the world taking direct, effective and inspirational action to address global climate and environmental crisis. While some of these children and young people have gained a public and media profile like Appadurai, Cullis-Suzuki and Thunberg, many work in obscurity, participating in local climate and environmental actions, far from the spotlight of global climate change negotiations. The media regularly run stories to introduce readers to "other Gretas" who are, at a young age, stepping up to take action to address the climate crisis: "Climate Kids: Meet the Global Network of Young Activists in Greta Thunberg's Green Army," "Meet the Teens Schooling Us on Climate," "Meet the British Greta Thunbergs Who Shoot From the Lip," "It's Not Just Greta Thunberg: Why Are We Ignoring the Developing World's Inspiring Activists?," "19 Youth Climate Activists You Should Be Following on Social Media" and so on (Galvez-Robles, 2019; Holthaus, 2018; Miller, 2019; Unigwe, 2019; Webster, 2020). All of this is fully in keeping with a long history of children and youth participating in movements for social change, including the civil rights, women's, labor and antiwar movements (Coles, 1986; Elshtain, 1996; Rodgers, 2020). Such child and youth political engagements swell up and come into prominence during key mobilizing periods of history, but are an ever-present part of social life.

Second, despite this history, there remains a sense of novelty and amazement that children and young people are speaking out and demanding action to tackle climate and environmental crises: it is not just what is said by individuals such as Appadurai, Cullis-Suzuki and Thunberg, but their young age that attracts attention and wonder. Alongside this sense of novelty comes a reaction of breathless hyperbole from many commentators—as seen in the dubbing of Cullis-Suzuki as "the Girl Who Silenced the World"—and exaggerated claims of the extraordinary powers of these young activists. We now see headlines such as "Will the Kids Save Us From Climate Change?," "The Climate Kids Are Coming: With a Green New Deal and Student Strikes for Climate, Will Young People Save Us Yet?," "Can They Save Us? Meet the Climate Kids Fighting to Fix the Planet," and "Will the Climate Kids Save the Planet?" (Grisé, 2019; Hertsgaard, 2019; Horton et al., 2019; Turrentine, 2018). There is no doubt that the speeches and actions of young leaders such as Appadurai, Cullis-Suzuki, Thunberg and others are inspirational. But—and this is a third key point raised by this brief history—there are also reasons to be cautious and ask careful questions about such hyperbolic reporting. For one thing, children and youth as a social group have the least amount of political and economic power of any age group in society: so what might it say about the state of climate crisis action if this is the group being delegated the task of taking lead on this action? For another, during the quarter century that stretches from Cullis-Suzuki's impassioned speech in Rio in 1992 to Thunberg's scolding address in Katowice in 2018, the climate crisis has gotten steadily worse. Over half of all CO2 emissions in the two hundred and seventy years since 1751 have been emitted in the thirty years since 1990 (Stainforth, 2020). Rates of global resource extraction, a principal cause of climate change and biodiversity loss, accelerated during this same period (Watts, 2019b). At the very least, these patterns should give pause to the often exaggerated nature of claims of the power of children and youth to change the world to fix the climate crisis.

This chapter explores claims made about the potential of children and young people to change society to address the climate crisis, and the empirical evidence of the nature, extent and impacts of child and youth climate activism. It argues that to think through the actual and potential roles that children and young people are playing and could play in climate activism, there is a need to reflect critically on dominant ideologies of childhood and climate action or agency; and acknowledge both the possibilities and limitations of child and youth power in relation to broader social and economic structures. There is also a need to look beyond a focus on children and youth, and consider the relationships and roles that adults and older generations have developed in support of politically engaged children and young people. This includes consideration of the role that education is playing and could play in fostering the development of more powerful young voices on the climate crisis, like those of Appadurai, Cullis-Suzuki and Thunberg.

Children and Youth Are Agents of Change!

There is a repeated mantra in climate change documents that children and youth are, could or should be "agents of change." UNICEF consistently promotes this rhetoric, arguing that "environmentally aware and empowered children and adolescents are potentially the greatest agents of change for the long-term protection and stewardship of the earth" (2007, p. 18). The Children in a Changing Climate (2015, p. 6) coalition likewise claims that children are "powerful agents of change, active at all levels, from the community, national, to international arenas." At the twenty-fifth annual United Nations Climate Change Conference in Madrid in 2019, ten countries signed an Intergovernmental Declaration of Children, Youth and Climate Action, to acknowledge, yet again, that children and young people play a "critical role as agents of change" (CERI, 2019).

How should we understand this mantra? It is notable that we rarely see any parallel claims being made about adults as agents of change. In part, what is going on here is a very important claim that challenges dominant stereotypes or norms of children and youth in contemporary western societies as being passive, apolitical and in a developmental state of becoming, thus lacking the competence and completeness that adults allegedly have. "Adults are usually assumed to possess agency, or to have the capacity to critically interpret the world and to act creatively upon that world," writes Jessica Taft (2019), "while children are frequently assumed to be passive objects of socialization, fully produced and constrained by their social contexts." Children's exclusion from the political process is typically based on these developmental assumptions and on a "presumption that childhood is a carefree period of innocence and that this state should not be disrupted by asking children to share in responsibility for the difficult aspects of contemporary life" (Moran-Ellis & Sünker, 2018, p. 281). What the "agents of change" mantra is essentially asserting is that, contrary to these norms, many children and young people are capable of taking on active roles to address the climate crisis, they want to take on these roles, they are being directly impacted in their own lives and communities by the climate crisis (and thus can hardly be said to be living in a pristine, protected state of innocence), and they should have full rights to take on such roles. As Youth vs. Apocalypse (2020), a youth-led climate justice group in the San Francisco Bay Area, insists, the principle is that "youth have the ability and right to be leaders in the fight for climate justice."

This claim for political inclusion of and respect for children and youth comes out of a discourse of children's rights. UNICEF links its call to recognise children and youth as agents of change directly to the United Nations Convention on the Rights of the Child (UNCRC):

> Article 12, together with the other key civil rights in the UNCRC, recognises the child as an active agent in the exercise of his or her rights. This right of active engagement has been broadly conceptualised as 'participation.'… Participation can be defined as an ongoing process of children's expression and active involvement in decision-making at different levels in matters that concern them. It requires information-sharing and dialogue between children and adults based on mutual respect, and requires that full consideration of their views be given, taking into account the child's age and maturity. (Lansdown, 2011, p. 3)

The call for political inclusion of children and youth in actions to address the climate crisis is also based on a principle at the heart of democratic education. Democratic education starts from the conviction that learning democracy is a central role of education in any democratic society, and that democracy must be learned experientially, through hands on practice and participation. If we take as a working definition of democracy Erik Olin Wright's (2010, p. 8) formulation that a democratic society is one in which "all people should have broadly equal access to the necessary means to participate meaningfully in decisions over things which affect their lives," then the claim of democratic education is that children and youth should be able to participate meaningfully in shaping collective decisions in the governance of their own schools certainly, but also in wider society (Apple & Beane, 2007; Moran-Ellis & Sünker, 2018). "Surely it is an obligation of education in a democracy," writes Maxine Greene (1985, p. 4), "to empower the young to become members of the public, to participate, and play articulate roles in the public space." One consequence of this, Dan Sabia (2012, p. 383) notes, is that democratizing education "means replacing the widespread idea that schools should prepare 'children for their future roles as citizens' with recognition that they are already 'citizens-in-fact.'"

This claim made by democratic education and children's rights discourse is a vital and radical claim, that runs counter to the ways in which schools and education systems are currently organized—as well as most political systems. It also directly challenges hegemonic visions of children as lacking in competence and capability. In democratic education, as Alex Bloom, founder of a democratic secondary state school in London in the mid-twentieth century, says, "what is needed most of all by teachers is a larger faith in the natural fitness of the child and in his [*sic*] inner potential" (quoted in Sabia, 2012, p. 381). But to the degree that the agents of change claim is based in a vision of radical, participatory and inclusive democracy, it should be clear what is being stated here: it is not that children and youth who step up as climate crisis activists possess some preternatural or supernatural ability to change the world that adults (and possibly "normal" children) lack. Indeed, this would be an odd claim for anyone committed to a vision of radical democracy to make. Rather, it is the insistence that children and youth have the ability and right to act on and transform the world around them, just like individuals of all other ages and life stages.

A second key question that needs to be asked of the children and youth as agents of change discourse is what kind of agency is being claimed or envisioned here? Invocations of child and youth participation are often framed as a secondary and highly circumscribed form of participation, so that, for example, rather than being given the right to vote, children are offered chances to participate in mock youth parliaments and assemblies (Moran-Ellis & Sünker, 2018). There is also a common pattern of elite (adult) actors drawing a line between good and acceptable (non-disruptive) versus bad and unacceptable (disruptive) forms of child and youth political action (Kennelly, 2011; Sukarieh & Tannock, 2008). These patterns show up in the realm of climate change action as well. UNICEF suggests that children can be agents of change through "promoting and adopting more sustainable low-carbon lifestyles" and "setting an example for their communities" (Narasimhan & Das, 2018; UNICEF, 2020). On the internet, we can find educational programs seeking to enable children to be "agents of change" by teaching them to do such things as care for others, conserve water, reduce, reuse, recycle and compost, practice empathy, tolerance and respect, improve their diet, and engage in more outdoors, physical activity (Buchbinder, 2018; White, 2020).

None of these aims are necessarily bad or wrong; however, they also make clear that children and youth are often being positioned as "agents of change" in a highly limited and individualized way, primarily through changing themselves: their own values, attitudes and behaviors. The suggestion is that "saving the planet becomes a lifestyle choice" and "small, individual acts of environmental consciousness model good behavior for others, who will be inspired to act accordingly": thus, children and youth can "be productive agents of change without engaging in difficult political struggle" (Maniates, 2016, p. 142). This is precisely the dominant focus on individual behavior change that has long been criticized as inadequate to tackling the root problems of global climate and environmental crisis (Maniates, 2001). More than this, as Karen O'Brien, Elin Selboe and Bronwyn Hayward (2018, p. 42) observe, youth climate activism can take on a range of different forms, from "dutiful" through "disruptive" to "dangerous dissent," and not all of these are received in a positive manner by "those with political power" or "strong interests in maintaining the status quo." While children and youth engaged in local recycling, energy conversation or school gardening initiatives may be widely celebrated, in the United Kingdom, for example, children who choose to go on strike in support of climate activism find themselves censured by government officials. "Every child should be in school," declared UK Secretary of Education Gavin Williamson in response to the mass school climate strikes that took place in September 2019: "They should be learning, they shouldn't be bunking off, and it's very irresponsible for people to encourage children to do so" (BBC, 2019). Children and youth who participate in actions with Extinction Rebellion in the UK—an environmental movement that uses nonviolent civil disobedience to pressure the government to take action on the climate crisis—risk being labelled by the state as extremists and reported to the police under UK anti-terrorism law (Dodd & Grierson, 2020; Holmwood, 2020).

Apart from changing themselves, another way in which children are frequently positioned in climate change discourse as "agents of change" is their potential to influence their parents' behaviors with respect to the climate crisis. At its best, the framework of what is called child-to-adult intergenerational learning turns dominant assumptions about childhood and education upside down, through providing opportunities for children and youth to become teachers and leaders, while adults take on the role of learners. This model has a long, rich history in radical education. In the 1961 Cuban literacy campaign, thousands of young, mostly female high school students were released from school early to teach adults basic literacy across the country, in what remains one of the most successful literacy campaigns in history (Herman, 2012; Murphy, 2011). Similarly, the development of the 1964 Freedom Schools in Mississippi in the United States was based on a vision of "the student as a force for social change," in which young African-American students would learn a political curriculum and organizing skills they could take into participating in and leading campaigns for desegregation and community empowerment throughout the state (Hale, 2011, p. 331; see also Chilcoat & Ligon, 1998). More recently, it is the mass school climate strikes that have been the most obvious example worldwide of children and youth seeking to educate, inspire and push adults into taking action on the climate crisis.

But this is not how much of the literature on child-to-adult intergenerational learning about climate change is framed. Instead, this literature tends to be individualized, concerned with how each child can influence his or her family's consumer and life style choices. Danielle Lawson and colleagues, for example, conducted an experimental study with middle school children in North Carolina, finding that climate change learning by children in school influenced their parents to increase their overall level of "climate change concern," with notable impacts by daughters and in politically conservative families (Lawson et al., 2019). Jane Spiteri

(2020) found that children attending Eco-Schools in Malta were able to influence their parents to adopt more "pro-environmental actions," such as recycling or conserving electricity. This individualized focus means that children are being expected to influence their parents without support of shared participation in a broader institutional or movement base (such as the Cuban literacy campaign, Mississippi Freedom Schools, or global student climate strikes), which raises ethical and political questions about undue burdens being placed on children, and children's ability to navigate often highly unequal power relationships within families. Indeed, researchers involved with these studies themselves acknowledge such concerns (Lawson et al., 2018, p. 206; Spiteri, 2020, p. 64). Raviro Chineka and Keiko Yasukawa's (2020, p. 589) study of intergenerational child-to-parent climate change learning in Zimbabwe found "multiple cultural and historical barriers to children being able to influence their parents," and argues that "children do not always succeed in influencing change [within their own families] because adults continue to assume the key decision-making roles and children's agency is not always perceived as having a sound foundation" (see also Phoenix et al., 2017).

Some of the literature on children influencing parental behavior in relation to the climate crisis comes not from the field of education, but adapts old consumer marketing techniques of exploiting children's "pester power," the "ability of children to unrelentingly request/pressurise their parents into buying products or services" (Sharma, 2020). Corporations and marketing experts speak of hoping to harness children's "positive pester power" to pressure their parents into doing things like buying environmentally friendly products, conserving energy use or reducing air travel (Gilmour, 2020; *Marketing* Week, 2008; O'Neill & Buckley, 2018). Npower, an energy provider in the UK, runs a Green Schools Campaign and Climate Cops website for children, where it encourages children to "remind anyone drying their washing in the electric tumble dryer on a nice day to use the washing line outside instead" (Vaughan, 2009). "Since kids are often driving the purchase of environmentally friendly products, there is potential for brands to make the most of this either via sustainable initiatives or affiliations with environmental causes," writes Rebekah Yock (2019), from the global advertising agency network M&C Saatchi: "Kids want to avoid palm oil if it means they're saving orangutan habitats." These models generally show little interest in developing rich learning spaces and practices in the family or home setting, but seek to manipulate family dynamics, children and parents into adopting consumer practices and products that positive pester power marketers have decided are the required solutions to the world's climate crisis.

The children and youth as agents of change discourse also needs to be understood as a statement of symbolic power. Karen Dubinsky (2012, pp. 7, 11), in an article that has a tongue in cheek title—"How Babies Rule the World"—notes how "popular images of highly politicized children" have been prominent "in the iconography of nation states, political parties, and social movements for many decades," and are widely used to "draw attention … and perhaps melt stony hearts." Images of children are politically potent, Dubinsky suggests, because "children symbolize the sentimental, the common hope of humanity abstracted from adult-created social problems" (p. 10). Similarly, youth has long been embraced by groups across the political spectrum as a key ideological symbol that signifies a "sense of regeneration and transformation" and is used to "promote the desirability of social change, and package and sell new ideologies, agendas, practices and products" (Sukarieh & Tannock, 2015, pp. 5, 82). The fact that a child or young person is speaking, participating or leading a campaign on an issue such as the climate crisis can draw public and media attention, and make an occasion notable, in a way that it would not be if it were just another adult. The ideology and iconography of childhood and youth in contemporary society can be powerful resources for movements for radical social transformation to draw upon.

However, this symbolic power has long been double-sided: just as it can be used to draw attention and build support, it may also be used to marginalize and contain. "Children appear so that adults can act," argues Dubinsky (2012, p. 8), who points out that while children are often embraced as political subjects and symbols, they are regularly denied serious recognition as legitimate political actors. To the degree that a demand for social change is seen as being made by children, this can increase its likelihood of being dismissed, ignored or treated lightly as amusing entertainment by adult political and economic elites. Likewise, the framing of social movements or political campaigns as youth-led can lead to these being treated by elites as "an irrational, silly, or ill-conceived approach to politics and social change," that is based in the immaturity and impatience of youth, and work "to obscure, marginalize and undermine … [the] serious and real social, political, economic, cultural and ideological conflicts that lie at the heart of these movements" and campaigns (Taft, 2011, p. 69; Sukarieh & Tannock, 2015, p. 112). In the context of Germany, Bergmann and Ossewaarde (2020, pp. 286, 287) have documented what they describe as a persistent "ageist derogation of the youth climate movement" by German media, that produces "an image of illegitimate youth climate activism and a structural underestimation of young activists' abilities and capabilities of knowing and exerting their climate interests." One of the obstacles to youth climate activists realizing their goals, Bergmann and Ossewaarde conclude, is not just the intransigence of fossil fuel interests in contemporary society, but "paternalist" and "ageist hegemonic representations of young people" (p. 287).

On Collective Actions and School Climate Strikes

While much of the children and youth as agents of change discourse takes on an individualized and symbolic frame, this does not mean that children and youth have only sought to address the climate crisis in these ways. Over the past decade, growing numbers of children and young people across the world have sought to develop and participate in collective actions, and mobilize direct social, political and economic power, to address the deepening climate crisis. Most commonly, children and youth have done this in exactly the same ways as, and in direct collaboration with, adult climate activists. Together, children, youth, adults and pensioners have participated in climate action marches, protests, campaigns, petitions, blockades and occupations. But there have been at least two forms of collective climate action that have mushroomed recently that are linked to these groups through their common identity as school, college and university students. These are the school climate strikes that spread worldwide since 2018, and university campaigns for fossil fuel divestment that first emerged in the United States in 2011. Here, I focus on the phenomenon of school climate strikes (for analysis of the fossil fuel divestment campaigns, see Ayling & Gunningham, 2017; Curnow & Gross, 2017; Grady-Benson & Sarathy, 2015; Healy & Debski, 2017).

The recent wave of global school climate strikes are generally traced back to Greta Thunberg's decision to start a one person school strike outside the Swedish parliament in Stockholm in August 2018. Thunberg's aims were to pressure the Swedish government to take urgent action to address the climate crisis. While initially out on strike every day until the Swedish general election in September, Thunberg shifted to striking one day a week, on Fridays, dubbing this "FridaysForFuture" (Witt, 2020). Thunberg's strike action attracted support and participation from others, at first in Sweden and then globally. In March 2019, an estimated 1.6 million people in over 130 countries around the world participated in a one-day school climate strike (Bergmann & Ossewaarde, 2020); in September 2019, over 6.5 million people joined in two consecutive global school climate strikes (Stuart et al., 2020). School climate strikes continued worldwide through to the end of 2019, before the Covid-19 pandemic and global lockdowns of 2020 forced a suspension of mass public gatherings. While the exact demands of school climate strikers vary from one place to another, these typically focus on pressuring national governments to follow climate science recommendations and international climate policy agreements, and take immediate, substantive actions that address the climate crisis effectively (De Moor et al., 2020).

These strikes have attracted enormous amounts of public, media and academic attention, and are credited with a number of initial successes. Dana Fisher (2019, p. 430) argues that the global school climate strikes engaged "more active participants in democracy," which "will have substantial and important consequences around the world." Heejin Han and Sang Wuk Ahn (2020, p. 17) suggest that the movement "succeeded in raising the profile of climate change as a pressing global issue of the highest priority, enlisting broad social endorsement and prompting incremental policy changes on the parts of some states." Han and Ahn point to the examples of France accelerating its emissions reductions, Germany declaring a phase-out of coal mining, and the United Kingdom passing a law "requiring the country to eliminate its carbon footprint" (p. 15).

Perhaps inevitably, however, the strikes have also been criticized, as some have asked what capability and power children and youth have that would be able to impact the climate crisis. "The core rhetoric and claims of the #FridaysForFuture movement are underdeveloped and lack necessary nuance," complains Darrick Evensen (2019, p. 428). "Youth climate activists do not have much leverage against governments and fossil fuel companies that resist change," note Han and Ahn (2020, pp. 16, 17), and lack "the power to bring about immediate policy changes." The climate strike movement is "in danger of becoming just another feel-good spectacle," warns Angi Buettner (2020, p. 834), "that "ends up merely being 'just beautiful words and images and promises,' the exact-same thing for which Greta Thunberg … accuses the current climate change responses of the world leaders and policy makers."

The school climate strikes are too recent to be able to make definitive claims about their long term significance (Nissen et al., 2020). But it can be helpful to put these strikes in the context of a long history of striking students around the world. For school strikes are not a new phenomenon. As Steve Cunningham and Michael Lavalette (2016, p. 20) observe, "the school student strike has been in existence for as long as we have had compulsory education." Localized and relatively spontaneous student strikes, walkouts, sit-ins and occupations are an ever present part of school experience, as students take immediate action over issues of local concern—changes to school dress codes, unfair student discipline or the actions of a disliked school principal. But there have also been periodic waves of mass school student strike actions in countries around the world. In the UK, national school student strike waves occurred in 1889, 1911, 1968–1974, 1978–1980, 1985 and 2003 (Cunningham & Lavallette, 2016, p. 88). While school strikes and walkouts usually focus on issues within the education sector, they have also addressed more global concerns. In 2003, there were school strikes around the world to oppose the US-led war on Iraq (Cunningham & Lavallette, 2004; Guardian, 2003; Vries, 2003); and in March 2018, there was a National School Walkout organized across the United States to call for stricter gun control laws, in the aftermath of the shooting of 17 high school students and staff in Florida (Yee & Blinder, 2018). This gun control walkout was later credited by Greta Thunberg as the inspiration for her decision to strike over climate crisis concerns in Stockholm, just five months later in August 2018 (Watts, 2019a).

From this long history of school strikes, several key patterns emerge. Student strike waves rarely occur in isolation, but commonly erupt in periods of general social movement mobilization. As Cunningham and Lavallette (2016), pp. 93, 94) point out, school strike waves tend to be:

> located within broader waves of social protest. … Large economic and political confrontations generate confidence … that collective action can win. This confidence can spread to the most marginalised sections [of society]—including school students—who start to fight for their rights. In this sense school strikes are … an important component part of a more generalised period of contentious political conflict.

This phenomenon has often led adult elites to dismiss student strikers as naïve copycats and pale imitators of adult political actions (Taylor, 1994). But it actually suggests something more significant: school strikes are evidence of powerful learning going on among students, about critical social problems and strategies for addressing these problems. Sometimes this education and learning is direct and deliberate. The wave of secondary student strikes that swept across Nicaragua during the 1970s was fostered, in part, by two decades of grassroots organizing in schools by Sandinista university students and community organizers (Rueda, 2020). Likewise, school strikes by African-American students participating in the US civil rights movement in the 1940s and 1950s, school walkouts (or "blowouts") by Chicano students in the United States in the 1960s, and the school strike in Soweto in South Africa in 1976, all were preceded by sustained, strategic and radical educating and organizing work by community organizers, teachers, university students and others (Baker, 2011; Brown, 2016a; García & Castro, 2011). At other times, this learning may be more indirect. In the 1889 student strike wave in the UK, it seems that many student strikers lived in families and communities that were at the heart of the widespread radical labor organizing going on at the time, and picked up their strike skills and strategies at home and in the neighborhood (Cunningham & Lavallette, 2016; Taylor, 1994).

This history has implications for how the recent global school climate strikes should be understood. These strikes are regularly framed as children and youth stepping in to make up for adult failure to act on the climate crisis: "the climate strikes are an example of youth becoming politicized, rejecting adult inaction and demanding more from governments," write Curnow and Helferty (2019); "this movement is about young people demanding that adults take responsibility for safeguarding their future," argue Wahlström and his co-authors (2019). In climate strike reporting, Walker (2019) observes, children are imagined "as the saviours of the climate movement." This is also how Thunberg and other child and youth climate activists often articulate their demands. When Thunberg was striking outside the Swedish parliament, she handed out leaflets that said, "I am doing this because you adults are shitting on my future" (Crouch, 2018). It is a theme she refers too often. "We children are doing this to wake the adults up," Thunberg (2019, p. 63) told the UK House of Parliament in 2019. "We're asking adults to step up alongside us," she wrote in the *Guardian* newspaper (Thunberg et al., 2019). But this framing is self-evidently not the whole story. For Thunberg and other young climate strikers also repeatedly ask political leaders to follow the warnings of (adult) climate scientists: "Listen to them, because we [children] are just repeating what they are saying and have been saying for decades" (Thunberg, 2019, p. 42). In other words, it is not that all adults are inactive about the climate crisis, only some adults—notably, those in political and economic leadership positions. Likewise, while most student climate strikers are not involved with formal environmental organizations (Wahlström et al., 2019), many core student climate strike leaders and organizers do work closely with (adult-led) environmental organizations, social movements and political parties (Marquardt, 2020). In contrast to the popular "children lead while adults do nothing" frame, it seems more likely that the recent wave of school climate strikes are a testimony to decades of effective climate change organizing and education occurring throughout society: they might be best seen as an indication of the growing strength of a largely adult-led climate crisis movement, rather than its weakness or absence.

Historically, localized student strikes have often been able to secure quick wins for demands made of their own schools. In 2013, for example, a student walkout in Wales was able to get a student reinstated, after he had been removed from class for violating a school ban against shaved heads—even though the student had shaved his head to raise money for a cancer charity (BBC, 2013). But larger student strike waves that make structural demands for change, whether in the education system or society more generally, rarely manage to secure many immediate substantive wins. There are obvious reasons for this, in that school students lack political or economic power, and student mobilization is hard to sustain, as student leaders age out and graduate from school. At the same time, student strike waves often have longer term significance as being key moments in the development of broader social movements that eventually *are* able to secure real change. The Chicano school student walkouts in 1968 are widely seen as a vital moment in the building of a powerful and effective Chicano movement in the United States (García & Castro, 2011). The Soweto school student strike in 1976 is likewise often seen as a key moment in the long term development of anti-apartheid struggle in South Africa, and dismantling of the apartheid state less than two decades later (J. Brown Brown, 2016a). The 2006 secondary school student strikes and occupations in Chile are commonly pointed to as an essential starting point for a growing social movement against neoliberal capitalism in the country that has continued to develop to the current day (Bellei & Cabalin, 2013; Larrabure & Leiva, 2018).

There are a number of reasons why student strikes often play a key role in the development of longer term social transformation. Sometimes, it is the response of elites to student strikes that is decisive: it was the violent brutality and murder of striking students in Soweto by South African police that led to mass uprisings against the apartheid state. But there are also core educational reasons for this role. Mass student strikes tend to garner considerable public and media attention, as a direct challenge to adult and state authority, and the norms of childhood and schooling. As a consequence, they can become essential events for raising levels of public consciousness and concern over the issues that students are striking over. As Arita Holmberg and Aida Alvinius (2020, p. 80) write about Thunberg and the global school climate strikes, "when a child defies or violates socially constructed rules and conventions such as school attendance and [procures] followers all over the world, the action receives major media attention." The fact that *even* children are moved to take action on an issue such as the climate crisis can be a consciousness raising moment for the general public. Alongside this public pedagogy—through which child and youth led strikes can help to educate external public audiences—the experience of organizing and participating in a school strike or walkout can be a profoundly transformative experience for child and youth strikers. In the past, child and youth school strikers have often gone on to play leading roles in social movements as they have grown up into adulthood: this is another factor that can make school strike waves a key starting point for social movement mobilization in the years and decades that follow. It would not be surprising to see similar developments transpire in the future with some of the recent child and youth climate strikers as well.

Behind Every Great (Young) Climate Activist...

One of the consequences of dominant stereotypes of children as being non-political, passive and lacking full developmental competence and capability is that when children do emerge as political activists there is a tendency to portray these children as exceptional and extraordinary (Brown, 2016b; Taft, 2020; Vanner & Dugal, 2020). The leadership, passion and strong voice of these children apparently come from nowhere other than some inner special ability these children are miraculously blessed with. Conversely, if evidence surfaces that adults have been working in collaboration with these political children, suspicion arises that these children are simply a front, subjects of adult manipulation, "pawns of adult activists," a manifestation of fraud, cheating and insincerity (Taft, 2019). An onus is placed on political children (and their adult allies) to show they are thinking and acting completely of their own accord. As Jessica Taft (2019) notes, defenders of "children's activism ... sometimes bend over backward to argue that these spaces are entirely child-led and express children's autonomous and authentic perspectives," and "downplay adult involvement." Greta Thunberg (2019, pp. 35–36), for example, is led to insist that her climate activism is entirely separate from and independent of any other climate change actors:

> Many people love to spread rumours saying that I have people 'behind me' or that I'm being 'paid' or 'used' to do what I'm doing. But there is no one 'behind' me except for myself. ... I am not part of any organization. ... I am absolutely independent and I only represent myself. ... I do what I do completely for free. ... And yes, I write my own speeches.

Taft (2020, pp. 11, 12) points out how this framing undermines effective child and youth activism, for it erases and makes invisible the "collective social movements" and "larger youth activist communities" that support the development and success of most if not all individual child and youth activists. It also undermines our ability to develop effective forms of education that can facilitate the social transformations necessary to address the climate crisis, as it suggests that education, whether formal, nonformal or informal, has played no role in the emergence of dynamic child and youth activists, and should properly play no such role either. As Lyn Mikel Brown (2016b, pp. 17, 20) argues, this myth of the "special" or "exceptional" child or youth activist strips away all of the "relationships, family stories, cultural background, lived experience" that provide the "scaffolding" that enables the success of each young activist. This is not only nonsensical, but deeply harmful. We need to always ask: what kinds of education and learning experiences shaped these powerful young actors, and how can we learn from this, to extend this education more widely? Children's activism, as Taft, Brown and others argue, "can and should be intentionally cultivated by caring, engaged, critical adults who want to foster young people's political power" (2019). To do this, we need to ask what lies behind each child and youth climate activist, not to undermine or cast aspersion on these young activists, but on the contrary, to extend our collective ability to foster rich, radical and transformative educational experiences for all children, youth and, indeed, adults.

Anjali Appadurai's educational background provides an excellent example of what can be gained by asking this question. The child of Indian immigrants to Canada, Appadurai won a scholarship to attend a United World College for an International Baccalaureate (Singh-Joseph, 2012). United World Colleges are a global network of residential schools that were set up in the 1960s with the aim of bringing students from all kinds of backgrounds around the world to study and live together, and a commitment to using education as a force for social change, that can help "unite people, nations and cultures for peace and a sustainable future" (UWC, 2020). The Colleges place strong emphasis on experiential learning, community service and fostering a "sense of idealism," "respect for the environment" and "intercultural understanding" (UWC, 2020; see also Tsumagari, 2010). After graduation, Appadurai won another scholarship to the College of the Atlantic in Maine for her undergraduate degree. This is a small alternative college that was created in 1969 "to focus on the relationship between humans and the environment," through the study of human ecology (College of the Atlantic, 2020). When Appadurai went to the United Nations Conference in Durban in 2011, she was part of a student delegation from this college (Davis UWC Scholars, 2012). Finally, in Durban, Appadurai was able to access the Conference of Youth, a three day program of workshops, trainings and networking opportunities with other young activists, that had been created several years earlier, in Montreal in 2005, to increase the presence and power of youth voice in the annual Climate Change Conferences (Appadurai, 2014; AYCC, 2011; Yona et al., 2020). In Appadurai, we thus have one great young climate activist and three very interesting and potentially important examples of alternative education projects. None of this takes away from Appadurai's own excellence as a public speaker and activist. But it opens the possibility of developing more powerful young voices like Appadurai through the expansion of these kinds of alternative educational spaces. The question can be extended: what other kinds of alternative education models can we look to that might offer ideas for how best to support the development of dynamic child and youth climate change activism?

For some young climate activists, teachers and lessons in mainstream state schools have played a key role in triggering their climate activism. "I was 10 and I had this amazing 5th grade social studies teacher who brought in stuff about real world problems into our lectures," says Haven Coleman of Denver, who co-founded US Youth Climate Strike at the age of thirteen: "I love, love, love sloths, so when I learned from him that deforestation endangered the sloth, I was like, 'I gotta do something.'" "While I was researching how I could stop deforestation, I found out about climate change and I was devastated," Coleman explains, "But, I didn't stop and just get sad, I felt like I needed to do something since nobody had been doing anything" (Minutaglio, 2019). For all the challenges of getting effective climate change education into the school curriculum, there is evidence that, for some students at least, this can make a difference. Overall, however, formal schooling does not appear to be a major source of inspiration for young climate activists. Indeed, one of the demands of school climate strikers in the United Kingdom is that "the education system is reformed to teach young people about the urgency, severity and scientific basis of the climate crisis" (UK Student Climate Network, 2020). Many schools do not just lack extensive climate change education; as Lynda Dunlop and her co-authors (2020, p. 4) point out, they can also present "regimes of obstruction to climate crisis activism" that "operate at different levels (classroom, school, curriculum)." Young climate activists are more likely to point to three other sources of learning for their climate activism: the influence of parents and family; the direct experience of harm inflicted by the climate crisis; and the example and inspiration of other young climate activists.

Kallan Benson, the fifteen-year old coordinator for Fridays for Future USA, is homeschooled. So is eleven-year old Elsie Luna, a UK striker with Fridays for Future and co-founder of Extinction Rebellion Kids. So is Xiuhtezcatl Martinez, now twenty, who for the past several years has been the youth director for Earth Guardians, a youth climate justice organization. So is Avery Tsai, a ten year old climate activist with Fridays for Future and Extinction Rebellion NYC. As is Dylan D'Haeze, sixteen years old, who has produced a series of documentary films called *Kids Can Save the Planet* (*No Kill* Magazine, 2020; Turns, 2019). Homeschooling, in itself, doesn't tell us much, as it can be shaped by radically different pedagogies and ideologies (Gaither, 2009). But we can ask what homeschooling practices and experiences were important for these young climate activists. Kallan Benson points not just to the influence of her parents, both of whom are environmental activists, but her experience of participating in homeschool programs run by the Smithsonian Environmental Research Center, and learning in Quaker meetings about Quaker traditions of activism (Minogue, 2020; Ross, 2019). Elsie Luna was home educated by her parents in the tradition of unschooling. "If I became interested in something, I was allowed to pursue it," writes Luna, as "my parents have supported me in what I want to do and learn" (Luna & Mearman, 2020, p. 6). Luna also points to the influence of her mother's environmental activism, and the opportunity to engage with and learn from communities of activist adults in her home as a young child. "When I was a baby, my mother set up a club at home for people who wanted to change their lives to fight against climate change," Luna explains, "they had get-togethers to chat." When Luna became active with Extinction Rebellion, she did so with the support of an 83 year old activist who "I have known … for a while," since he "was part of the group my mother set up when I was young" (Luna & Mearman, 2020, p. 6). Xiuhtezcatl Martinez speaks of the importance of being able to learn traditional Aztec ideas from his father at home—"that sense of caretaking of the planet, that everything around us is a gift, and we have to protect it"—and traditions of activism, power and passion from his mother—"her ability to put everything aside and to fight for what's right" (McPherson, 2015).

While Severn Cullis-Suzuki wasn't homeschooled, she had a remarkable educational experience growing up at home. It is not just that Cullis-Suzuki's parents are well known environmentalists in Canada, but "their house in Vancouver … always attracted intellectuals and activists who want to change the world" (McIlroy, 2003). When Cullis-Suzuki was young, a Kayapo family from the Amazon forest came to stay with her family after the father received death threats in Brazil for leading protests against plans to build a hydroelectric dam; when the family was able to return to Brazil, Cullis-Suzuki and her parents went to stay with them. She thus was able to have a direct, transformative experience of the struggles of Indigenous communities to protect their forest environment in the Amazon, simply through being her parents' child (Stainsby, 1993; Wilson, 2014). Kelsey Juliana, the named plaintiff in a children's climate justice lawsuit against the United States government, likewise, was not homeschooled. But her parents, environmental activists themselves, strongly influenced her decision to become a climate activist, taking her as a young child "to rallies and marches in their hometown of Eugene, Oregon" and using "their bodies physically against logging lorries" to stop clear cuts in the Oregon forest (Maynard, 2019; Powell, 2018). Ridhima Pandey, a twelve year old climate activist in Utterakhand in northern India, has similarly been able to learn from her father, who is also an environmental activist, and mother, who is a forestry guard (Farand, 2017).

The role of parents and family in shaping and supporting brilliant young climate activists, in other words, is often vital, even as this role comes in different shapes and sizes. Sometimes, parents themselves may not be environmental or climate activists, but still provide their children with essential encouragement, opportunity and support. Greta Thunberg's parents seem to have gone to extraordinary lengths to enable their daughter's development as one of the world's best known climate activists (Ernman et al., 2020). This central role of parents raises a question about why so much attention is being paid by some climate change researchers to seeking to mobilize children to transform their parents' climate and environmental behaviors at home. We cannot, of course, choose who our parents are: this is a matter of luck. But, as suggested in the following section, we could pay a lot more attention to ways of supporting the learning, well-being and practice of parents directly, with the idea that their children, too, are likely to benefit, rather than seek to do this in roundabout ways through their children. "The climate crisis is a parenting problem," argues Mary DeMocker (2018) in *The Parents' Guide to Climate Revolution*, and "parents are key to saving the planet."

Direct experience of local harms inflicted upon home, family and community by the climate crisis is also widely cited by young climate activists as a fundamental starting point for their activism. "I grew up in California, where climate change has been around me since I was little," says Alexandria Villaseñor, the fifteen year old co-founder of US Youth Climate Strike: "One year, we had this massive drought where our lakes dried up completely" and "I started saying, 'This is not normal and this doesn't make me feel safe anymore'" (Minutaglio, 2019). In 2018, on a return visit to California with her family, Villaseñor, who is asthmatic, "was caught in the cloud of smoke from the Camp Fire"—the "most destructive and deadliest wildfire in California history," that killed 85 people—and "for days afterward she felt physically ill and emotionally distraught" (Hagerty, 2019; Kaplan, 2019). Ridhima Pandey in India was moved to take climate action not just by her parents but after her "whole family was displaced by the Uttarakhand floods of 2013, which claimed hundreds of lives'; for Kaluki Paul Mutuku, a young climate activist in Kenya, it was the growing impact of climate change induced drought that spurred him to take action, as he saw his mother and other mothers in his village having to walk further and further just to get water (Unigwe, 2019). Greta Thunberg's decision to launch her school strike in August 2018 was spurred, in part, by Sweden having its hottest summer in recorded history and suffering from "the country's 'most serious' wildfire situation of modern times" (Crouch, 2018; The Local, 2018). A 2019 feature report on the Sunrise Movement in the US found that almost every Sunrise member interviewed had "a story about how climate dramatically made a mark on their lives"—including record hurricanes, flooding and wildfires (Nilsen, 2019).

As Naomi Klein (2019, p. 4) argues, for children and young people in many parts of the world, climate crisis is increasingly something learned firsthand, as "climate disruption on a planetary scale is not a future threat, but a lived reality." The importance of direct experience of climate crisis in fostering climate activism lends support to arguments made by Katharine Hayhoe and others, explored in Chap. 7, of the key role that self interest will likely play in moving large numbers of people to take the kind of actions that can effectively address this crisis. It also reinforces the arguments about place that were discussed in Chap. 5. The most effective educational projects to address the climate crisis may need to start with children and young people in those countries, regions and neighborhoods that are being most significantly impacted by the climate crisis, right here and right now. As the climate crisis deepens, of course, this group of children and young people is becoming ever larger by the day.

Once child and youth climate activism takes hold, it can have a rapid snowball effect, as seen in the remarkable spread of the FridaysForFuture school climate strike movement around the world, from the moment Greta Thunberg began her school strike in Stockholm in August 2018 through to the arrival of the Covid-19 pandemic at the end of 2019. In social movement theory, this is commonly discussed as a phenomenon of diffusion of ideas and tactics, that can occur rapidly during cycles of protest (Soule, 2004; Walsh-Rousso, 2014). But it is also a process of education and learning, as "activists are … inspired and guided by other activists" (Haydu, 2020, p. 626). One argument made by social movement researchers about diffusion is that activist identity is often central to these processes. "When protesters see themselves as *like* protesters elsewhere," writes Jeffrey Haydu (2020, p. 626), "facing comparable problems, perhaps sharing a common identity, they may emulate movement practices seen from afar." What is clear in the talk and writing of young climate activists is how effective and important it can be for children and young people to learn ideas and strategies of climate crisis activism from each other. In learning and taking inspiration from the actions of someone like Greta Thunberg, it is not just her specific ideas and actions that matter to children and youth, but the fact that she is a young child, just like they are. "She is 16, the same age as me," says Howey Ou, a young climate activist in southwestern China, of being inspired to act by Thunberg: "She had such a deep recognition of the climate crisis," and "I then thought that I could also be like her starting from now, to use a huge amount of passion and courage to do what I believe is right" (Chen, 2019). "It makes me want to get involved," says a high school senior in the US, "seeing someone the same age as me, or even younger than me, doing something so big and making a difference makes me want to make a difference too and shows me that I can" (Mackin, 2019). "When Greta started striking, it gave permission to other students all around the world to go on protest as well," says Alexandria Villaseñor (Bendix, 2019).

This kind of peer-to-peer, mutual learning is a central principle and practice in many traditions of progressive, democratic, feminist and radical education; and yet is all too often under-developed and even obstructed in (adult) teacher-centered models that prevail in much mainstream schooling. Furthermore, even if it is the case, as Catherine Walker (2020) suggests, that most child and youth climate activism around the world is likely being done quietly, anonymously and on an everyday basis, the importance of learning from the example of other young climate activists suggests that work done by some of the most high profile child and youth climate activists, even if relatively unusual and non-representative, can be invaluable in providing ideas and inspiration to other children and young people, some of whom may live in very distant and different social and environmental contexts, to develop their own forms of climate crisis action as well.

What About the Adults?

In 1959, philosopher Hannah Arendt published a strong critique of the political use of children in the US civil rights movement. Focusing on a confrontation over desegregation of education that took place at Little Rock Central High School in Arkansas—where fifteen year old African American student Elizabeth Eckford was prevented from entering by members of the Arkansas National Guard, along with a mob of white protesters—Arendt (1959, p. 50) condemned the move "to burden children … with the working out of a problem which adults for generations have confessed themselves unable to solve." As she chastized Eckford's "absent father" and "the equally absent representatives of the NAACP," Arendt posed the question: "Have we now come to the point where it is the children who are being asked to change or improve the world?" (p. 50).

Arendt's critique was controversial from even before the moment it was published, and has been the focus of ongoing contestation and debate ever since. One of the enduring concerns raised by Arendt is that allowing children to become directly involved with political struggles can be harmful to their development—which, as discussed in Chap. 4, arises in the context of the climate crisis with worries about the potentially damaging effects of eco-anxiety. The radical, democratic and progressive education response to such concerns, as noted earlier, is to seek to ensure that education about social problems always begins with the needs, interests and experiences of the individual learner, so that it fits their age, life stage and social situation; and also to note that sometimes "*not* going public when the world is in turmoil may be much harder on children than for them to take up their share of the burden" (Elshtain, 1996, p. 20). Another strong response to Arendt's critique is to assert the political agency of children: to argue that children often have the capability and interest in taking political actions, childhood and education are always already political, whether we wish them to be or not, and children should have the right to take political action, as democratic citizens alongside adults (e.g., Elshtain, 1996; Kallio, 2009; Nakata, 2008; Schutz & Sandy, 2015).

Arendt's critique of the incident at Little Rock also raises questions regarding the position of adults when children take political action, as she claims there is an abrogation of duty and responsibility. One response to such criticism is to show that, actually, adults are usually closely involved whenever children and young people act, whether in the context of the US civil rights movement decades ago or the climate justice movement today (Sukarieh & Tannock, 2015). Kathryn Gines (2014, pp. 16–19) shows how Arendt was wrong about the absence and lack of responsibility of Elizabeth Eckford's parents and NAACP (National Association for the Advancement of Colored People) representatives: both her parents and NAACP representatives, along with many other adults, were closely involved in supporting and assisting Eckford and the other children who were trying to desegregate Little Rock Central High School. As noted above, it is often stereotypes and prejudices about childhood that lead to a downplaying of adult involvement in the actions of political children. As a consequence, supporters of children's activism call for more recognition and development of effective, caring, equitable and supportive intergenerational activism to address endemic social problems such as the climate crisis (Brown, 2016b; Taft, 2019).

There is, however, another respect in which the "what about the adults" question needs to be addressed. In discussions about climate change education, the focus often is on children and young people: what is being done, what could and should be done, in primary and secondary schools, colleges and universities to teach about the climate crisis. When Robert Stevenson, Jennifer Nicholls and Hilary Whitehouse (2017, p. 68) pose the question "what is climate education?," their answer is "climate change education involves creatively *preparing children and young people* for a rapidly changing, uncertain, risky and possibly dangerous future" (emphasis added). Likewise, in David Rousell and Amy Cutter-Mackenzie-Knowles' (2020, p. 192) "systematic review of climate change education," the focus is solely on children and young people, because they are "future leaders whom the public expects to overcome the legacies of environmental inaction," and "who will be forced to grapple with the uncertain effects of climate change brought forth by previous generations." There is a concerning stereotype of adults in some discussions of climate change education as poor, ineffective or unwilling climate change learners. In some research on using children's to get parents to become more concerned about climate change, for example, the claim is made that children rather than adults are "key in spurring climate action" because adult learners are (allegedly) often "anti-reflexive," in a way that "clouds their judgement when forming perceptions on controversial subjects," are less able than children "to parse scientific fact from political contexts," and hold opinions driven by "political ideologies and worldviews" (Lawson et al., 2018, p. 205).

In contrast to this focus on children and youth, one of the arguments made by Myles Horton, founder of Highlander Folk School, is that if we want to develop forms of education that can help to facilitate radical social transformation, it is adult education that is essential, more so than the education of children and young people:

> The idea of the Highlander Folk School … was to try to use adult education as one of the main mechanisms for changing society. I had come to see that it was wrong for adults to always say: 'The younger generation is going to change society,' and then for them to go ahead and fix it so that it would be impossible for the young to do just that. I decided that if you're going to do anything about changing society—through education—it has to be with adults. (Horton, 2003, p. 11)

The argument made here is demographic and pragmatic: most of the population in most countries are not children but adults, and adults have more social, political and economic power than children. It is also temporal, as the suggestion is that social problems need to be dealt with here and now, not left to future generations to solve. Child and youth climate change learning and activism is excellent and encouraging; but if we are serious about using education to support radical social change to address the climate crisis, we need to focus on developing adult education as well. Since adults are usually not primarily students, but are engaged in doing other things such as working, parenting and caring, this means needing to develop nonformal and informal education that is easy to access and drop in and out of (Derrick, 2020). As Jennie Stephens and Amanda Graham (2008, p. 204) note, in the United States, only about 20% of the total population (mostly children and youth) are enrolled in formal education, and "formal educational mechanisms, processes, and institutions do not reach 80% of the population, people whose decisions, actions, and attitudes have significant implications for society's response to climate change."

If we return to the example of the US civil rights movement and longer traditions of African American freedom struggles through education, while there have been outstanding examples of children and young people taking a central role—in the confrontation over school desegregation in Little Rock in 1957, the Children's Crusade in Birmingham, Alabama in 1963, or the Mississippi Freedom Schools in 1964—the development of alternative, informal and nonformal forms of adult education was central. This includes Highlander Folk School, as a residential school that brought together local community leaders from across the US South for the explicit purpose of learning how to enact radical social change (Horton, 2003). It includes the Citizenship Schools developed by Septima Clark and others, that mobilized hundreds of community educators and thousands of adult students across the US South to teach and learn basic literacy skills, alongside a program of political education (Levine, 2004). It also includes the adult education programs and workshops on black history that were developed by Carter Woodson and his Association for the Study of Negro Life and History with the aim of fostering "an awakening of the masses" (Woodson, 1933; see also Apple, 2013).

One of the major concerns and challenges for developing effective climate change education has nothing to do with children or young people, or schools or universities. It is that we live in an era when adult education is widely neglected and ignored. UNESCO's (2019, p. 12) *Fourth Global Report on Adult Learning and Education* found that "adult learning and education remains low on the agenda of most Member States—participation is patchy, progress inadequate and investment insufficient." This is particularly the case for what UNESCO refers to as "liberal, popular and community education" that focuses on the development of "active citizenship skills" (p. 38)—the kind of education that addresses issues like the climate crisis. What limited adult education does exist tends to be vocational, oriented to developing workforce knowledge and skill. In the UK, the Centenary Commission on Adult Education (2019, pp. 2, 27) raised concerns that adult participation in learning activities has been declining, as "during the past decade of 'austerity', the country's educational and cultural infrastructure has deteriorated." This includes not just the direct provision of adult education classes, but also the loss of "high-quality public facilities, including libraries, museums, art galleries, theatres, music venues, parks, and community centres," that can support non-formal and informal learning opportunities for adults (p. 27).

Both UNESCO and the UK Centenary Commission argue that adult education is a "permanent national necessity," vital not just for supporting a healthy democratic society, but specifically for addressing the challenge of climate crisis (Centenary Commission on Adult Education, 2019, p. 4). "Our quality of life in the future will depend on averting a climate catastrophe," states the Commission on Adult Education (2019, p. 51): "If any justification is required for providing increased educational opportunities for all, surely this is it—the importance of enabling educated discussion of such grave issues and problems, and how to go about tackling them." "Put adult learning and education at the centre of your efforts to achieve sustainable economies and societies and recognize its key role in developing integrated, holistic solutions to the problems we face," UNESCO (2019, p. 19) concurs: "At this moment in time, nothing less will do." An article by Jana Ahlers about the child and youth led Fridays for Future school strike movement perhaps sums up the answer to the "what about the adults" question as effectively as anyone. "No matter how much one admires the revolutionary potential of the youth" who are "on the streets today," Ahlers (2020) writes, "they cannot by themselves shape the political and social structures that are necessary for a green and just transition." Fridays for Future is important, argues Ahlers, but "we need Adults for Future! as well." Many young climate activists say the same. "Young people have led the climate strikes," Greta Thunberg and forty-six other young climate activists wrote in the *Guardian* newspaper in May 2019: "We need adults to join us too" (Thunberg et al., 2019). "Politicians always tell me, 'Keep up the good work; your generation is going to save the world!,'" writes Zero Hour founder Jamie Margolin (2020, p. 139) in her book, *Youth to Power*: "They are both so right and so unbelievably wrong at the same time." "To make changes in our world," Margolin insists, "our movements must be intergenerational" (p. 15).

References

Ahlers, J. (2020). Adults for future! *Electronic platform for adult learning in Europe*. 23 September. https://epale.ec.europa.eu/en/blog/adults-future

Appadurai, A. (2014). Earth in brackets. In J. Foran, C. Ellis, & S. Gray (Eds.), *At the COP: Global climate justice youth speak out* (pp. 15–23). International Institute of Climate Action & Theory.

Apple, M. (2013). Can education change society? Du Bois, Woodson and the politics of social transformation. *Review of Education, 1*(1), 32–56.

Apple, M., & Beane, J. (Eds.). (2007). *Democratic schools: Lessons in powerful education*. Heinemann.

Arendt, H. (1959). Reflections on Little Rock. *Dissent, 6*(1), 45–56.

AYCC—Australian Youth Climate Coalition. (2011). Conference of Youth. *YouTube*. https://www.youtube.com/watch?v=k7dOIkTceLE

Ayling, J., & Gunningham, N. (2017). Non-state governance and climate policy: The fossil fuel divestment movement. *Climate Policy, 172*, 131–149.

Baker, S. (2011). Pedagogies of protest: African American teachers and the history of the civil rights movement, 1940–1963. *Teachers College Record, 113*(12), 2777–2803.

BBC. (2013). Protest over Milford Haven school charity head shave boy. *BBC News*, 4 October. https://www.bbc.co.uk/news/uk-wales-south-west-wales-24398589

BBC. (2019). Climate strike: Thousands protest across UK. *BBC News*, 20 September. https://www.bbc.co.uk/news/uk-49767327

Bellei, C., & Cabalin, C. (2013). Chilean student movements: Sustained struggle to transform a market-oriented educational system. *Current Issues in Comparative Education, 15*(2), 108–123.

Bendix, A. (2019). Greta Thunberg isn't the only teenage climate activist you should know. *Business Insider*, 7 December. https://www.businessinsider.com/greta-thunberg-youth-climate-movement-activist-backlash-2019-12?r=US&IR=T

Bergmann, Z., & Ossewaarde, R. (2020). Youth climate activists meet environmental governance: Ageist depictions of the FFF movement and Greta Thunberg in German newspaper coverage. *Journal of Multicultural Discourses, 15*(3), 267–290.

Brown, J. (2016a). *The road to Soweto*. James Currey.

Brown, L. (2016b). *Powered by girl*. Beacon Press.

Buchbinder, D. (2018). Three ways to help kids become agents of change. *TwentyThirty*, 16 March. https://twentythirty.com/article/three-ways-kids-become-agents-of-change/

Buettner, A. (2020). 'Imagine what we could do'—The school strikes for climate and reclaiming citizen empowerment. *Continuum, 34*(6), 828–839.

Camerini, V. (2019). *Greta's story: The schoolgirl who went on strike to save the planet*. Aladdin.

Centenary Commission on Adult Education. (2019). A permanent national necessity…. In *Adult education and lifelong learning for 21st century Britain*. University of Nottingham.

CERI—Children's Environmental Rights Initiative. (2019). Declaration on children, youth and climate action. https://www.childrenvironment.org/declaration-children-youth-climate-action

Chen, L. (2019). Meet the Greta Thunbergs of China fighting climate change. *Goldthread*, 12 November. https://www.goldthread2.com/identity/greta-thunberg-china-climate-change/article/3037404

Chilcoat, G., & Ligon, J. (1998). 'We talk here. This is a school for talking': Participatory democracy from the classroom out into the community: How discussion was used in the Mississippi Freedom Schools. *Curriculum Inquiry, 28*(2), 165–193.

Children in a Changing Climate. (2015). *Realising children's rights in a changing climate*. Children in a Changing Climate.

Chineka, R., & Yasukawa, K. (2020). Intergenerational learning in climate change adaptations: Limitations and affordances. *Environmental Education Research, 26*(4), 577–593.

Coles, R. (1986). *The political life of children*. Atlantic Monthly Press.

College of the Atlantic. (2020). Mission & history. https://www.coa.edu/about/mission-history/

Crouch, D. (2018). The Swedish 15-year-old who's cutting class to fight the climate crisis. *Guardian*, 1 September. https://www.theguardian.com/science/2018/sep/01/swedish-15-year-old-cutting-class-to-fight-the-climate-crisis

Cullis-Suzuki, S. (1993). *Tell the world: A young environmentalist speaks out*. Doubleday Canada.

Cunningham, S., & Lavallette, M. (2004). 'Active citizens' or 'irresponsible truants'? School student strikes against the war. *Critical Social Policy, 24*(2), 255–269.

Cunningham, S., & Lavallette, M. (2016). *Schools out! The hidden history of Britain's school student strikes*. Bookmarks.

Curnow, J., & Gross, A. (2017). Injustice is not an investment: Student activism, climate justice, and the fossil fuel divestment campaign. In J. Conner & S. Rosen (Eds.), *Contemporary youth activism* (pp. 367–382). Praeger.

Curnow, J., & Helferty, A. (2019). A year of resistance: How youth protests shaped the discussion on climate change. *The Conversation*, 19 December. https://theconversation.com/a-year-of-resistance-how-youth-protests-shaped-the-discussion-on-climate-change-129036

Davis UWC Scholars. (2012). A young voice challenges the climate summit. https://www.davisuwcscholars.org/stories/node/5471

De Moor, J., et al. (2020). New kids on the block: Taking stock of the recent cycle of climate activism. *Social Movement Studies*. https://doi.org/10.1080/14742837.2020.1836617

DeMocker, M. (2018). *The parents' guide to climate revolution*. New World Library.

Democracy Now! (2011). 'Get it done': Urging climate justice, youth delegate Anjali Appadurai mic checks UN summit. 9 December. https://www.democracynow.org/2011/12/9/get_it_done_urging_climate_justice

Derrick, J. (2020). Adult learning infrastructure is a central feature of a healthy democracy. *FE News*, 17 January. https://www.fenews.co.uk/fevoices/224-podcasts/40721-adult-learning-infrastructure-is-a-central-feature-of-a-healthy-democracy-dr-jay-derrick-ucl

Dodd, V., & Grierson, J. (2020). Terrorism police list Extinction Rebellion as extremist ideology. *Guardian*, 10 January. https://www.theguardian.com/uk-news/2020/jan/10/xr-extinction-rebellion-listed-extremist-ideology-police-prevent-scheme-guidance

Dubinsky, K. (2012). Children, ideology, and iconography: How babies rule the world. *Journal of the History of Childhood and Youth, 5*(1), 7–13.

Dunlop, L., et al. (2020). The role of schools and teachers in nurturing and responding to climate crisis activism. *Children's Geographies*. https://doi.org/10.1080/14733285.2020.1828827

Elshtain, J. (1996). Commentary: Political children. *Childhood, 3*, 11–28.

Ernman, M., et al. (2020). *Our house is on fire: Scenes of a family and a planet in crisis*. Penguin.

Evensen, D. (2019). The rhetorical limitations of the #FridaysForFuture movement. *Nature Climate Change, 9*, 428–433.

Farand, C. (2017). Nine-year-old girl files lawsuit against Indian government over failure to take ambitious climate action. *Independent*, 1 April. https://www.independent.co.uk/environment/nine-ridhima-pandey-court-case-indian-government-climate-change-uttarakhand-a7661971.html

Fisher, D. (2019). The broader importance of #FridaysForFuture. *Nature Climate Change, 9*, 428–433.

Foran, J., Gray, S., & Grosse, C. (2017). 'Not yet the end of the world': Political cultures of opposition and creation in the global youth climate justice movement. *Interface, 9*(2), 353–379.

Fridays for Future. (2020). Strike statistics. https://fridaysforfuture.org/what-we-do/strike-statistics/

Gaither, M. (2009). Homeschooling in the USA: Past, present and future. *Theory and Research in Education, 7*(3), 331–346.

Galvez-Robles, I. (2019). 19 youth climate activists you should be following on social media. *EarthDay.org*, 14 June. https://www.earthday.org/19-youth-climate-activists-you-should-follow-on-social-media/

García, M., & Castro, S. (2011). *Blowout! Sal Castro & the Chicano struggle for educational justice*. University of North Carolina Press.

Gilmour, H. (2020). Connecting with the Alphas—The brand power behind the pester power. *WARC*, 14 April. https://www.warc.com/newsandopinion/opinion/connecting-with-the-alphas%2D%2Dthe-brand-power-behind-pester-power/3535

Gines, K. (2014). *Hannah Arendt and the negro question*. Indiana University Press.

Goodman, A. (2011). On climate change, the message is simple: Get it done. *Guardian*, 14 December. https://www.theguardian.com/commentisfree/cifamerica/2011/dec/14/durban-climate-change-conference-2011

Grady-Benson, J., & Sarathy, B. (2015). Fossil fuel divestment in US higher education: Student-led organising for climate justice. *Local Environment, 21*(6), 661–681.

Greene, M. (1985). The role of education in democracy. *Educational Horizons, 63*, 3–9.

Grisé, C. (2019). Will the climate kids save the planet? *Upfront*, 18 February. https://upfront.scholastic.com/issues/2018-19/021819/will-the-climate-kids-save-the-planet.html#1300L

Guardian. (2003). Pupils walk out over war. 5 March. https://www.theguardian.com/uk/2003/mar/05/politics.schools

Hagerty, C. (2019). The Camp Fire, one year later. *Vox*, 16 October. https://www.vox.com/the-highlight/2019/10/16/20908291/camp-fire-wildfire-california-paradise-survivors

Hale, J. (2011). 'The student as a force for social change': The Mississippi Freedom Schools and student engagement. *Journal of African American History, 96*(3), 325–347.

Han, H., & Ahn, S. (2020). Youth mobilization to stop global climate change: Narratives and impact. *Sustainability, 12*(10), 1–23.

Haydu, J. (2020). Adding time to social movement diffusion. *Social Movement Studies, 19*(5), 625–639.

Healy, N., & Debski, J. (2017). Fossil fuel divestment: Implications for the future of sustainability discourse and action within higher education. *Local Environment, 226*, 699–724.

Herman, R. (2012). An army of educators: Gender, revolution and the Cuban Literacy Campaign of 1961. *Gender & History, 24*(1), 93–111.

Hertsgaard, M. (2019). The climate kids are coming: With a Green New Deal and student strikes for climate, will young people save us yet? *Nation*, 28 January. https://www.thenation.com/article/archive/greta-thunberg-climate-change-davos/

Holmberg, A., & Alvinius, A. (2020). Children's protest in relation to the climate emergency: A qualitative study on a new form of resistance promoting political and social change. *Childhood, 27*(1), 78–92.

Holmwood, J. (2020). Ratcheting up the Prevent agenda in schools. *Discover Society*, 4 November. https://discoversociety.org/2020/11/04/ratcheting-up-the-prevent-agenda-in-schools/

Holthaus, E. (2018). Meet the teens schooling us on climate. *Grist*, 27 February. https://grist.org/article/meet-the-teens-schooling-us-on-climate-change/

Horton, M. (2003). *The Myles Horton reader*. University of Tennessee Press.

Horton, A., McClinton, D., & Aratani, L. (2019). Can they save us? Meet the climate kids fighting to save the planet. *Guardian*, 4 March. https://www.theguardian.com/environment/2019/mar/04/can-they-save-us-meet-the-climate-kids-fighting-to-fix-the-planet

Jaud, J. (2010). *Severn: The voice of our children*. Wide Management.

Kallio, K. (2009). Between social and political: Children as political selves. *Childhoods Today, 3*(2), 1–22.

Kaplan, S. (2019). How a 7th-grader's strike against climate change exploded into a movement. *Washington Post*, 16 February. https://www.washingtonpost.com/national/how-a-7th-graders-strike-against-climate-change-exploded-into-a-movement/2019/02/15/e20868e2-2fb4-11e9-86ab-5d02109aeb01_story.html

Kennelly, J. (2011). *Citizen youth: Culture, activism, and agency in a neoliberal era*. Palgrave.

Klein, N. (2019). *On fire: The burning case for a Green New Deal*. Penguin.

Lansdown, G. (2011). *Every child's right to be heard*. Save the Children.

Larrabure, M., & Leiva, F. (2018). The rise of a new left in Chile? *NACLA*, 19 January. https://nacla.org/news/2018/01/19/rise-new-left-chile

Lawson, D., et al. (2018). Intergenerational learning: Are children key in spurring climate action? *Global Environmental Change, 53*, 204–208.

Lawson, D., et al. (2019). Children can foster climate change concern among their parents. *Nature Climate Change, 9*, 458–462.

Levine, D. (2004). The birth of the Citizenship Schools: Entwining the struggles for literacy and freedom. *History of Education Quarterly, 44*(3), 388–414.

Luna, E., & Mearman, A. (2020). Learning to rebel. *Sustainable Earth, 3*(4), 1–10.

Mackin, E. (2019). Global climate strike: The youth have spoken. *The Spotlight*, 31 October. https://slspotlight.com/our-world/2019/10/31/6938/

No Kill Magazine. (2020). Avery is an activist. https://www.nokillmag.com/articles/avery-is-an-activist

Maniates, M. (2001). Individualization: Plant a tree, buy a bike, save the world? *Global Environmental Politics, 1*(3), 31–52.

Maniates, M. (2016). Make way for hope: A contrarian view. In S. Nicholson & S. Jinnah (Eds.), *New earth politics: Essays from the Anthropocene* (pp. 135–154). MIT Press.

Margolin, J. (2020). *Youth to power*. Hachette.

Marquardt, J. (2020). Fridays for Future's disruptive potential: An inconvenient youth between moderate and radical ideas. *Frontiers in Communication, 5*, 1–18.

Maynard, M. (2019). Generation change. *Geographical*. April Issue, 18–25.

McIlroy, A. (2003). Don't call her an environmentalist: At 12 years old, David Suzuki's daughter moved the world to tears at the Earth Summit. *Globe and Mail*. 1 February: F5.

McPherson, C. (2015). Meet the teenage indigenous hip-hop artist taking on climate change. *Rolling Stone*, 13 July. https://www.rollingstone.com/politics/politics-news/meet-the-teenage-indigenous-hip-hop-artist-taking-on-climate-change-186079/

Miller, B. (2019). Climate kids: Meet the global network of young activists in Greta Thunberg's green army. *ABC News*, 30 September. https://www.abc.net.au/news/2019-10-01/climate-change-young-activists-fight-for-environment-action/11556360

Minogue, K. (2020). Student activist Kallan Benson on standing up for climate change. *Shorelines*, 14 April. https://sercblog.si.edu/volunteer-spotlight-student-activist-kallan-benson-on-standing-up-for-climate-change/

Minutaglio, R. (2019). The world is burning: These girls are fighting to save it. *Elle*, 14 March. https://www.elle.com/culture/career-politics/a26755814/youth-climate-strike-march-15/

Moran-Ellis, J., & Sünker, H. (2018). Childhood studies, children's politics and participation: Perspectives for processes of democratisation. *International Review of Sociology, 28*(2), 277–297.

Murphy, C. (2011). *Maestra*. http://www.maestrathefilm.org

Nakata, S. (2008). Elizabeth Eckford's appearance at Little Rock: The possibility of children's political agency. *Politics, 28*(1), 19–25.

Narasimhan, G., & Das, M. (2018). Protecting and empowering children. *Climate2020*, 9 November. https://www.climate2020.org.uk/protecting-and-empowering-children/

Nilsen, E. (2019). The new face of climate activism is young, angry—And effective. *Vox*, 17 September. https://www.vox.com/the-highlight/2019/9/10/20847401/sunrise-movement-climate-change-activist-millennials-global-warming

Nissen, S., Wong, J., & Carlton, S. (2020). Children and young people's climate crisis activism—A perspective on long-term effects. *Children's Geographies*. https://doi.org/10.1080/14733285.2020.1812535

O'Brien, K., Selboe, E., & Hayward, B. (2018). Exploring youth activism on climate change: Dutiful, disruptive and dangerous dissent. *Ecology and Society, 23*(3), 42–54.

O'Neill, C., & Buckley, J. (2018). 'Mum, did you just leave that tap running?!' The role of positive pester power in promoting sustainable consumption. *International Journal of Consumer Studies, 43*(3), 253–262.

Part, M. (2019). *The Greta Thunberg story: Being different is a superpower*. Sole Books.

Phoenix, A., Boddy, J., Walker, C., & Vennam, U. (2017). *Environment in the lives of children and families*. Policy Press.

Powell, A. (2018). Meet the kids trying to put the government on trial for its climate policies. *Pacific Standard*, 29 October. https://psmag.com/environment/meet-the-kids-trying-to-put-the-government-on-trial-for-its-climate-policies

Rodgers, D. (2020). *Children in social movements: Rethinking agency, mobilization and rights*. Routledge.

Ross, J. (2019). Giving up childhood to avert climate catastrophe. *Friends Journal*, 1 December. https://www.friendsjournal.org/giving-up-childhood-to-avert-climate-catastrophe/

Rousell, D., & Cutter-MacKenzie-Knowles, A. (2020). A systematic review of climate change education: Giving children and young people a 'voice' and a 'hand' in redressing climate change. *Children's Geographies, 18*(2), 191–208.
Rueda, C. (2020). A La Huelga! Secondary students, school strikes, and the power of educational activism in 1970s Nicaragua. *The Americas, 77*(4), 601–631.
Sabia, D. (2012). Democratic/utopian education. *Utopian Studies, 23*(2), 374–405.
Schutz, A., & Sandy, M. (2015). Friendship and the public stage: Revisiting Hannah Arendt's resistance to 'political education'. *Educational Theory, 65*(1), 21–38.
Sharma, A. (2020). Changing behaviors with the new pester power for good. *Medium*, 12 July. https://medium.com/@anannya.sh/changing-behaviors-with-the-new-pester-power-for-good-206a305c8dae
Singh-Joseph, R. (2012). Spotlight: Anjali Appadurai. *Darpan*, 1 February. https://www.darpanmagazine.com/people/newsmakers/anjali-appadurai/
Smith, I. (1993). All in the family: Severn Suzuki, 13-year-old daughter of environmentalist David Suzuki, is following in the footsteps of her activist parents. *Vancouver Sun*, 8 August: C12.
Soule, S. (2004). Diffusion processes within and across movements. In D. Snow, S. Soule, & H. Kriesi (Eds.), *Blackwell companion to social movements* (pp. 294–310). Blackwell.
Spiteri, J. (2020). Too young to know? A multiple case study of child-to-parent intergenerational learning in relation to environmental sustainability. *Journal of Education for Sustainable Development, 14*(1), 61–77.
Stainforth, T. (2020). More than half of all CO2 emissions since 1751 emitted in the last 30 years. *Institute for European Environmental Policy*, 29 April. https://ieep.eu/news/more-than-half-of-all-co2-emissions-since-1751-emitted-in-the-last-30-years
Stainsby, M. (1993). The kid who could: At the age of 13, Severn Cullis Suzuki is following in the eco footsteps of her activist parents. *Vancouver* Sun, 10 July: B8.
Stead, E., & Baker, L. (2020). *Planet Greta: How Greta Thunberg wants you to help her save our planet*. Scholastic Children's Books.
Stephens, J., & Graham, A. (2008). Climate science to citizen action: Energizing nonformal climate science education. *Eos, 89*(22), 204–205.
Stevenson, R., Nicholls, J., & Whitehouse, H. (2017). What is climate change education? *Curriculum Perspectives, 37*, 67–71.

Stuart, D., Gunderson, R., & Petersen, B. (2020). The climate crisis as a catalyst for emancipatory transformation: An examination of the possible. *International Sociology, 35*(4), 433–456.

Sukarieh, M., & Tannock, S. (2008). In the best interests of youth or neoliberalism? The World Bank and the new global youth empowerment project. *Journal of Youth Studies, 11*(3), 301–312.

Sukarieh, M., & Tannock, S. (2015). *Youth rising? The politics of youth in the global economy*. Routledge.

Taft, J. (2011). *Rebel girls: Youth activism and social change across the Americas*. New York University Press.

Taft, J. (2019). *The kids are in charge: Activism and power in Peru's movement of working children*. New York University Press.

Taft, J. (2020). Hopeful, harmless, and heroic: Figuring the girl activist as global savior. *Girlhood Studies, 13*(2), 1–17.

Taylor, T. (1994). As the old cocks crow the young ones learn: The school strikesof 1889 and the New Union movement. *History of Education, 23*(1), 89–106.

The Local. (2018). What you need to know about Sweden's historic wildfire outbreak. 17 July. https://www.thelocal.se/20180717/sweden-battles-most-serious-wildfire-situation-of-modern-times-heres-what-you-need-to-know

Thunberg, G. (2019). *No one is too small to make a difference*. Penguin.

Thunberg, G., et al. (2019). Young people have lead the climate strikes: Now we need adults to join us too. *Guardian*, 23 May. https://www.theguardian.com/commentisfree/2019/may/23/greta-thunberg-young-people-climate-strikes-20-september

Tsumagari, M. (2010). The enduring effects of a United World College education as seen through a graduate's eyes. *Journal of Research in International Education, 9*(3), 289–305.

Tucker, Z., & Persico, Z. (2019). *Greta and the giants*. Frances Lincoln Children's Books.

Turns, A. (2019). Meet generation Greta: Young climate activists around the world. *Guardian*, 28 June. https://www.theguardian.com/environment/2019/jun/28/generation-greta-young-climate-activists-around-world

Turrentine, J. (2018). Will the kids save us from climate change? *Natural Resources Defence Council*, 23 March. https://www.nrdc.org/onearth/will-kids-save-us-climate-change

UK Student Climate Network. (2020). We, the students, demand…. https://ukscn.org/our-demands/

UNESCO. (2019). *Fourth global report on adult learning and education*. UNESCO.
UNICEF. (2007). *Climage change and children*. UNICEF.
UNICEF. (2020). Environment and climate change: Climate change and environmental degradation undermine the rights of every child. https://www.unicef.org/environment-and-climate-change
Unigwe, C. (2019). It's not just Greta Thunberg: Why are we ignoring the developing world's inspiring activists? *Guardian*, 5 October. https://www.theguardian.com/commentisfree/2019/oct/05/greta-thunberg-developing-world-activists
UWC—United World Colleges. (2020). What is UWC? https://www.uwc.org/about
Vanner, C., & Dugal, A. (2020). Personal, powerful, political: Activist networks by, for, and with girls and young women. *Girlhood Studies, 13*(2), vii–xv.
Vaughan, A. (2009). Children aren't kidding when it comes to green issues. *Guardian*, 30 January. https://www.theguardian.com/environment/ethicallivingblog/2009/jan/30/carbon-emissions
Vries, L. (2003). Students cut class to protest war. *CBS News*, 10 March. https://www.cbsnews.com/news/students-cut-class-to-protest-war/
Wahlström, M., et al. (2019). *Protest for a future: Composition, mobilization and motives of the participants in Fridays for future climate protests on 15 March, 2019 in 13 European cities*. Centre for the Understanding of Sustainable Prosperity, University of Keele. https://www.cusp.ac.uk/themes/p/protest-for-a-future-report/
Walker, C. (2019). Victims, saviours or villains? Children in popular climate imaginaries. *Discover Society*, 6 February. https://discoversociety.org/2019/02/06/victims-saviours-or-villains-children-in-popular-climate-imaginaries/
Walker, C. (2020). Uneven solidarity: The school strikes for climate in global and intergenerational perspective. *Sustainable Earth, 3*(5), 1–13.
Walsh-Rousso, C. (2014). Diffusion of protest. *Sociology Compass, 8*(1), 31–42.
Watts, J. (2019a). Greta Thunberg, schoolgirl climate change warrior. *Guardian*, 11 March. https://www.theguardian.com/world/2019/mar/11/greta-thunberg-schoolgirl-climate-change-warrior-some-people-can-let-hings-go-i-cant
Watts, J. (2019b). Resource extraction responsible for half world's carbon emissions. *Guardian*, 13 March. https://www.theguardian.com/environment/2019/mar/12/resource-extraction-carbon-emissions-biodiversity-loss

Webster, B. (2020). Meet the British Greta Thunbergs who shoot from the lip. *Times*, 1 February. https://www.thetimes.co.uk/article/meet-the-british-gretas-who-shoot-from-the-lip-3656vdj7j

Marketing Week. (2008). Positives of pester power. 23 January. https://www.marketingweek.com/positives-of-pester-power/

White, H. (2020). Education for sustainability: Agents of change. *Early Years Educator*, 10 March. https://www.earlyyearseducator.co.uk/features/article/education-for-sustainability-agents-of-change

Wilson, J. (2014). *Severn and the day she silenced the world*. Second Story Press.

Witt, E. (2020). How Greta Thunberg transformed existential dread into a movement. *New Yorker*, 6 April. https://www.newyorker.com/books/under-review/how-greta-thunberg-transformed-existential-dread-into-a-movement

Woodson, C. (1933). *The mis-education of the Negro*. CreateSpace.

Wright, E. (2010). *Envisioning real utopias*. Verso.

Yee, V., & Blinder, A. (2018). National school walkout: Thousands protest against gun violence across the US. *New York Times*, 14 March. https://www.nytimes.com/2018/03/14/us/school-walkout.html

Yock, R. (2019). From the bottom up—The new pester power for good. *Marketing*, 2 September. https://www.marketingmag.com.au/hubs-c/opinion-yock-pester-power-good/

Yona, J., et al. (2020). Applying a leverage points framework to the United Nations climate negotiations: The (dis)empowerment of youth participants. *Elementa, 8*(36), 1–14.

Youth vs. Apocalypse. (2020). Points of unity. https://youthvsapocalypse.wixsite.com/website/points-of-unity

7

The Role of Self Interest

Katharine Hayhoe, an atmospheric scientist and climate communication leader, makes the case that until large numbers of people start to see climate change as affecting them directly and mattering in their lives right here right now, there will be little hope of substantive action capable of redressing the crisis. "The most pernicious and dangerous myth we've bought into when it comes to climate change is not the myth that it isn't real or humans aren't responsible," Hayhoe argues, "it's the myth that it doesn't matter to me" (Friedman, 2017). While Hayhoe (2016, 2019) acknowledges other motivations—concern with what is happening to others less fortunate than ourselves, or to the other species with whom we share this planet—she insists self interest is the crux of the matter. This "simple yet potentially revolutionary understanding," Hayhoe (2017) claims, means "getting people to care about a changing climate doesn't require adopting 'new' values," or "inspiring people to 'care' about deforestation and melting ice caps," but instead, "imparting urgency and concern is just a matter of showing people how to connect the dots among the issues they already care about" in their own lives. What is capable of teaching large numbers of people that taking action to address the climate crisis is in their own interests, regardless of personal identity, faith

S. Tannock, *Educating for Radical Social Transformation in the Climate Crisis*, Palgrave Studies in Education and the Environment,
https://doi.org/10.1007/978-3-030-83000-7_7

or ideology? In part, this is learning to communicate and educate effectively, in ways that help make the link between the crisis and people's interests. But in part, it is the encroachment of the crisis itself, as it comes to have ever more severe impacts closer to more and more people's own homes. The reason there has been an upsurge in concern and activism over the climate crisis in North America and Europe in recent years is principally because the effects of the crisis are no longer only obvious in far off places, affecting distant peoples, animals and lands, but are hitting hard here as well, with record-breaking floods, wildfires, hurricanes and droughts. The big question for Hayhoe is when the "tipping point" will come that will "motivate sufficient action" to address the climate crisis—what Hayhoe calls the "collective 'oh shit!' moment, when people finally realise climate impacts do pose a far greater threat than the solutions" (Watts, 2019). Will this moment come soon enough to limit the worst effects of the growing climate and environmental crisis, or will it come when we are already well on our way to seeing worst case future scenarios come true?

In making these claims, Hayhoe is, in some ways, saying no more than what most social justice approaches to education also claim: learning is most powerful when it is centered around problems that learners recognize as meaningful and relevant to them personally. But Hayhoe is also pushing a more pragmatic, realpolitik line than what many radical, progressive, social justice education and organizing approaches tend to embrace, as these are more likely to call for a shifting of values, ideologies, principles and priorities, and development of compassion, empathy, solidarity, love and caring for others. Although there is no direct link, Hayhoe is essentially embracing an argument that has been controversial in social justice education and organizing, and is associated with Saul Alinsky and groups who embrace an Alinskyite approach to community organizing: this is to focus on the importance of self interest in generating effective collective social action. Self interest, as Alinsky (1971, p. 53) writes in *Rules for Radicals: A Pragmatic Primer for Realistic Radicals*, "wears the black shroud of negativism and suspicion," and "is associated with a repugnant conglomeration of vices such as narrowness, self-seeking, and self-centeredness, everything that is opposite to the virtues of altruism and selflessness." Alinsky argues, however, that "the myth of altruism as a motivating factor in our behavior" is "one of the classic American fairy tales," and insists that "to question the force of self-interest that pervades all areas of political life is to refuse to see man [*sic*] as he is, to see him only as we would like him to be" (pp. 53–54). The issue of self interest is often overlooked and avoided in social justice and climate change education. But it is essential to bring to the fore, as it is one of the most important issues that needs to be addressed for developing any kind of effective climate change education. What a discussion of self interest does is bring a concern with *the political* into the mix of educating and organizing around the climate crisis—a crisis that has often been positioned, peculiarly and unhelpfully, as outside and beyond the realm of politics altogether, in what some scholars criticize as a prevailing discourse in the contemporary period of the "post-political."

Climate Change Education: Post-Political Discourse and Win-Win Framing

Dominant approaches to dealing with the climate crisis have tended to be marked not only by technocratic and managerialist discourse, but a positioning of the crisis as being above and beyond the realm of politics. Sometimes, this is done directly and explicitly. Kenis and Lievens (2014, p. 531) quote former UK Prime Minister Tony Blair's claim that "global warming is too serious for the world any longer to ignore its danger or split into opposing factions on it"; and sociologist (and Blair adviser) Anthony Giddens' argument that "climate change should be lifted out of a right-left context, where it has no place," since "the issue is so important and all-encompassing that the usual party conflict should be suspended or muted." But it is also done through discursive moves or framings that some theorists term the "post-political," claimed to be hegemonic in contemporary neoliberal society (MacGregor, 2014). Indeed, Erik Swyngedouw (2010, p. 216) argues that climate change is one of the "key arenas through which the post-political frame is forged, configured and entrenched."

The climate crisis is often presented as a "universal humanitarian threat" in which all humans are "potential victims" (Swyngedouw, 2010, p. 221). In the face of this "common existential threat," the "material and ideological differences between people" are deemed to fade in significance, since "when we are all in the same leaking boat careening toward the apocalypse, there is no space, time, or need for politics" (MacGregor, 2014, p. 620). The climate crisis is also often said to have been caused by the actions of all humans everywhere—as in the claim that we are currently living in the Anthropocene epoch, a period in which nature is being extensively reshaped by human activity in general. "Every human being who has ever lived has played a part in making us the dominant species on this planet," argues Amitav Ghosh (2016, p. 115), "and in this sense every human being, past and present, has contributed to the present cycle of climate change" (quoted in Malm & Warlenius, 2019). In light of this universal human responsibility, climate change action focuses on the need to shift individual environmental behaviors: if we are each responsible for causing the crisis, the solution is that we each need to focus on changing what we ourselves are doing. Once again, radical differences in power, wealth and material interests between individuals and groups are shunted into the background. We are asked to "pledge to change our [own] behavior," writes MacGregor (2014, p. 621), "rather than question the global and local asymmetries and inequities that create, sustain, and legalize institutional forms of environmental exploitation." Finally, in the face of this overwhelming crisis, "scientific expertise becomes the foundation and guarantee for properly constituted politics/policies," and key decisions and choices in virtually "all aspects of life" are "increasingly considered to be a question of expert knowledge and not of political position" (Swyngedouw, 2010, pp. 217, 225).

Environmental and climate change education have also been marked by this apolitical approach (Henderson & Zarger, 2017). Much climate change education is framed as science education, focusing on increasing climate change understanding among students through "conveying factual information about climate science" (Monroe et al., 2019, p. 2): it thus does not even seek to address the political education questions of "how to change the world," or how to "shift social and material conditions in ways that lead to a tangible decline in carbon emissions and toward increased forms of resilience and flourishing for both humans and the more-than-human world" (Henderson & Drewes, 2020, p. 2; Henderson, 2019). In part, this approach is reinforced by the expectation or even legal requirement in schools and other formal education settings that teachers should avoid taking overt political stances, in favor of ideals of classroom "objectivity" or "impartiality" (Stevenson, 2007). If climate change as a topic "becomes political," it may be viewed by some schools and educators as being "too close to advocacy for classroom educators to address" (Monroe et al., 2019, p. 2). In part, too, this approach is fostered by the continued prevalence of a (long discredited) cognitivist or rationalist theory of social change in environmental education, that presumes a process of increased "environmental knowledge leading to environmental awareness and concern (environmental attitudes), which in turn [leads] to pro-environmental behavior" (Kollmuss & Agyeman, 2002, p. 241). In this model, it is clarity of scientific knowledge that matters, rather than any close engagement with the messiness of social politics.

As discussed in Chap. 3, much climate change education focuses on fostering individual proenvironmental behavior change, which has become what Kenis and Mathijs (2012, p. 45) call "a kind of 'holy grail' to tackle climate change in environmental policy [and] the environmental movement." This "individualization of responsibility," as Maniates (2001, p. 33) and others point out, is depoliticizing, as it leaves "little room to ponder institutions, the nature and exercise of political power, or ways of collectively changing the distribution of power and influence in society." Climate change action is "effectively 'privatized'" (Jorgenson et al., 2019, p. 164). "Saving the planet becomes a lifestyle choice rather than a political act," writes Maniates, and environmentalism is construed "as an individual, rational, cleanly apolitical process that can deliver a future that works without raising voices or mobilizing constituencies" (2016, p. 142; 2001, p. 41). In parallel with this individualized, behaviorist approach, environmental and climate change education often promotes environmental learning as "a win-win process … across different social scales," such that this kind of "learning is … largely assumed to entail no losses" for any individual or group (Boström et al., 2018, p. 5). This is a world of social transformation that is somehow conflict and struggle free, helped along by managerialist interventions, technological breakthroughs and voluntary shifts in individual behavior. We are all expected to support and engage with climate change education, as we will all benefit handsomely from it.

There is a paradox in all of this. Despite the often intense, conflictual social politics that surround efforts to include climate change in the school curriculum, the actual ways in which climate change is taught and learned as an educational topic is commonly, in and of itself, apolitical or depoliticized (see Kenis, 2019 for a general discussion of this paradox within contemporary discourse of climate change as a whole).

The problem with such post-political or apolitical approaches is that they are not very helpful for understanding the nature of the climate crisis, nor supporting efforts that can successfully address the crisis. Universalist views of the climate crisis fail to recognise the "massively uneven" distribution of responsibility for causing the crisis. Malm and Warlenius (2019) point out that just 90 corporations are responsible for producing over 60% of all industrial carbon dioxide and methane emissions over the past 250 years worldwide. Consequently, they obscure what needs to be done and who needs to be confronted to address the crisis. Such views also fail to recognise the significance of the radical unevenness of current and likely future impacts of the climate crisis. Climate apartheid and climate barbarism, in which, as Naomi Klein (2019) writes, "toxic ideologies that rank the relative value of human lives" are unleashed "in order to justify the monstrous discarding of huge swaths of humanity," are an active threat on the planet, while human extinction as an outcome of the crisis remains extremely unlikely (despite claims of groups like Extinction Rebellion). As Boström et al. (2018, pp. 5, 14) point out, learning, in general, is "a 'political' process in which knowledge and skills such as convincing argumentation … are used to promote certain interests and values over others," and thus most if not all forms of learning will "almost necessarily produce both winners and losers"; the same may be said for social change overall, as "there is no conflict-free transformation." The issue of the climate crisis, more particularly, "is intrinsically political in that it expresses fundamental conflicts of interests, power and values" (Mangat et al., 2018, p. 189); and any transition to a radically different, more just and sustainable society will never be "simply a technological or indeed a socio-technical matter," but will need to be made through a "deeply political struggle" that is "characterized by issues of power, distribution of and access to resources, [and] political economy" (Healy & Barry, 2017, p. 452).

If we are to develop an effective climate crisis education, we need to acknowledge this inherently political nature of the crisis, and address the importance of power (discussed in Chap. 8), self interest and conflict in shaping the crisis, as well as responses to it. This means, as Katharine Hayhoe argues, addressing the self interest of individuals and communities in taking action to address the crisis. But a core element of taking a political as opposed to post-political approach to the climate crisis is to recognize that individuals and groups have different interests at stake, a different "something that matters" to them in relation to the crisis, and some of these interests will inevitably be irreconcilable (Van Poeck & Östman, 2018, p. 1408). The realm of the political is marked by conflict, dissensus and disagreement (Håkansson & Östman, 2019; Sund & Öhman, 2014). This means recognizing the self interest of some in *not* taking any action to address the climate crisis, in taking action to save themselves while excluding and abandoning others in relation to the crisis, or actively working to obstruct and undermine efforts to take action to resolve the climate crisis. There are always going to be winners and losers in the unfolding of the climate crisis and actions taken to address the crisis: the question of who these will and should be is not a scientific or technical question, but one that is inherently political. To address this political nature of the climate crisis and delve into the importance of self interest, there are clear traditions in radical education that are helpful to draw upon. The Alinskyite model of community organizing is one tradition; but the long history of theory and practice in democratic education and consciousness raising, and more recent work in critical pedagogy also have much to offer.

Alinsky, Community Organizing and the Importance of Self Interest

Self interest is a central concept in Alinskyite approaches to community organizing, in two distinct ways. One is the principle that to be effective in mobilizing a community to take collective action for social transformation, you need to start by finding out and working with the interests, issues and values that matter most to individuals, groups and institutions within that community. Coming to a community to tell people what they should care about is unlikely to be effective; nor is hoping to rely solely on community altruism to support the needs and interests of others elsewhere. People are more likely to take sustained, collective action to change something in society when they see it as in their interests (Chambers, 2018; Sen, 2003; Walls, 2015).

While community organizing is often left out of school-based discussions of social justice education, the process of identifying, understanding and taking action to support community interests is seen by Alinsky and other organizers as a form of "popular education," that Alinsky (1969, p. 155) argues is "the ultimate objective implicit within democracy." Popular education (in Alinsky's terms) starts by learning about the self interests of individuals in a local community. Self interest is understood in a broad sense, distinguished from selfishness ("what's in it for me?") and selflessness ("whatever you want"), to "encompass all of those aspects of people that motivate them to act," their "needs, values, and purposes," their "agenda, whether economic, social, or spiritual," their priorities "not simply about what someone wants to get, but about 'who' someone is or wants to be" (Hart, 2001, p. 66; Schutz & Sandy, 2011, p. 194; Walls, 2015, p. 64). For community organizers, self interest is understood to be a social concern: it is not just about "personal greed," but "includes issues a person cares about, the people they are most likely to listen to, how they feel about their reputation, which groups they belong to and value, and more" (Schutz & Sandy, 2011, p. 236). Community members go on to learn that "what they considered primarily their individual problem is also the problem of … others" in their community, and that different issues in the community are often related to one another, as the common "progeny of certain fundamental causes" (Alinsky, 1969, pp. 59, 156). Community members also learn to look beyond the local community, to develop an analysis of how "many apparently local problems are in reality malignant microcosms of vast conflicts, pressures, stresses and strains of the entire social order" (Alinsky, 1969, p. 60)—a matter, as Katharine Hayhoe puts it, of "connecting the dots." The principle of centering community organizing and education on local interests is one of democracy and strategy. Learning is something that people come to feel compelled to do "because of the learning itself which is essential to their own life" (Alinsky, 1969, p. 165).

An awful lot of climate change discourse in its post-political framing proceeds in virtually the opposite manner to the community organizing approach. An elite science led consensus has been formed that climate change is the single most important issue affecting all of humanity, more significant than any local political differences and grievances, and a question is then posed about why more people are not doing more to prioritize addressing this crisis. A "misanthropic" rhetoric of accusation and berating emerges, in which different flaws and deficits that allegedly benight humanity are identified (Malm & Warlenius, 2019): problems with short-term thinking, irrational behavior, or how "human brains are wired" (Marshall, 2014). As the authors of *A Planet to Win* put it, in this discourse, "we have met the enemy and he is us" (Aronoff et al., 2019, p. 38). As seen in Chap. 3, ever more clever ways to nudge people into acting in the correct ways are devised through expert research and experimentation. However, if we follow the community organizing argument that people are likely to be moved to act on issues, like the climate crisis, if and when they see these issues as impacting their own local interests, the converse approach might be what is actually needed. "I feel like climate activists are always frustrated people don't care about their issue," says Varshini Prakash, co-founder of the Sunrise Movement in the United States: "The problem is we're not listening to what people care about" (Nilsen, 2019). It is not enough to present scientific arguments about the importance of climate change: these need to be linked to the interests people have, and any collective action to address the climate crisis needs to address and support these interests as well.

In many ways, much of what has been discussed in the preceding chapters on the importance of developing a pedagogy of hope (Chap. 4) and pedagogy of place (Chap. 5) is precisely about linking the problem of the climate crisis to the situated interests, needs, values and aspirations of local communities. But we also see the embrace of this approach in recent developments like the proposed Green New Deal in the United States, that combines a climate action target of reducing carbon emissions in the country to zero, with a goal of eliminating the severe inequalities of race, class and gender that exist in the US, by calling for a job guarantee for everyone who wants one, universal, free access to higher education and health care, and provision of safe, healthy and affordable housing and food for all. Echoing the language of Alinsky, supporters of the Green New Deal insist there is "a practical case for radicalism" in these sweeping proposals (Aronoff et al., 2019, p. 17):

> Radical change only happens when millions of people are organizing, striking, and marching, shaping politics and the economy from below. Tackling the climate crisis will require actions from unions, social movements, Indigenous peoples and others.... That's why the Green New Deal must combine climate action with attacks on social inequality. Only then can we build enough public support and grassroots organizing to break the stranglehold of the status quo, and give people reasons to keep fighting for more. (Aronoff et al., 2019, pp. 6–7)

"People mobilize around concrete projects that appeal to their desires and values," write Kate Aronoff and her co-authors (2019, pp. 7, 173), and thus "a Green New Deal would ... have to make climate action viscerally beneficial" to the communities they hope to mobilize. Matt Huber (2019) makes a similar argument, warning that the environmental movement has often had a strong middle class bias "directly antagonistic to working-class interests." To build the mass mobilization that will likely be necessary to fight successfully for the kinds of radical social transformations needed to address the climate crisis, Huber argues that it is essential to develop an "ecological politics for the working class," through mobilizing "around environmentally beneficial policies that appeal to the material interests of the vast majority of the working class mired in stagnant wages, debt, and job insecurity."

The second way in which self interest is central to community organizing is the need to develop a clear analysis of "who the key players are on any given issue, what their interests are, and whether they will be friendly or hostile" to the aims of a community organizing campaign on this issue (Phulwani, 2016, p. 871). The question asked is: "whose interests would put them in opposition to you, and whose interests might make them allies?" (Walls, 2015, p. 78). Community organizing works in a world of political realism and rejects (post-political) assumptions that scientific evidence, logical argument or moral principle will be sufficient for winning everyone over to a cause (Phulwani, 2016). There will always be conflicting interests of individuals, groups and institutions and differential power in realizing these interests. In "the real world," as distinct from the world as we might like it to be, "democracy is dominated by the interests of a few wealthy and powerful institutions" and actors (Chambers, 2018, p. 2). To be effective, community organizing campaigns need to have clear targets ("who or what group has the ability to make the change you want"), and a keen understanding of who will be a campaign's strongest opponents (those whose interests stand directly against the social change demands being made by a campaign) (Schutz & Sandy, 2011, p. 235). Alinsky (1971, p. 131) points to the strategic importance for organizing in a "complex, interrelated, urban society," where elites are adept at displacing and obscuring their responsibility for creating social problems, in clearly "identifying the enemy," to be able "to single out who is to blame for any particular evil." The analysis of the identity, power and self interest of campaign targets and opponents is seen as a vital part of the popular education learning process.

In climate change education and organizing, we see this approach most clearly in the fossil fuel divestment movement, that calls for universities and other institutions to become "fossil free" by divesting from fossil fuel and rejecting fossil fuel industry sponsorship (Malm & Warlenius, 2019). The divestment movement emerged as a reaction against the post-political discourse that frames mainstream climate change policy (Healy & Barry, 2017; Kenis & Lievens, 2014). Rather than unjustly, inaccurately and unhelpfully blaming an amorphous body of humanity for causing the climate crisis, the movement is based on a careful analysis of who the most powerful actors are who bear the greatest responsibility for causing the crisis and obstructing action to address the crisis—namely, the fossil fuel industry—and what institutional targets can be pressured to cut their ties with this industry. Again echoing Alinsky, the movement's strategy is to "turn climate change from a complex global problem into a simple battle between good and evil," insist that "climate change can be understood as a struggle between competing social forces, and [demand] we take sides" (Cheon & Urpelainen, 2018, p. 10; Mangat et al., 2018, p. 190). Movement leaders argue that "identifying a tangible enemy is crucial to success" in organizing, and one problem for climate crisis organizing previously is that it "lacked a clearly defined adversary" and "didn't have a clear cut bad guy" (Gunningham, 2017, p. 378). While ostensibly the divestment movement is concerned with imposing financial pressure on fossil fuel corporations, its broader agenda is moral, educational and political, as it seeks to shift public consciousness around fossil fuels and climate change (Ayling & Gunningham, 2017): to "remove the 'social license' by which fossil fuel companies operate through reputational damage and stigmatization," and "delegitimize fossil fuels" (Healy & Barry, 2017, p. 453); and reframe "the climate change narrative from a technocratic analysis of carbon emissions to a human-centred narrative that calls for systemic social and economic change" (Healy & Debski, 2017, p. 715).

While Alinskyite community organizing is helpful in foregrounding a concern with the importance of conflicting self interests in developing campaigns to address the climate crisis and other social problems, it is not without its own limitations. For developing effective education in relation to the climate crisis, two limitations are particularly important. Alinsky (1971, p. 54) recognized that self interest could and often should be changed radically, as he wrote of the "shifting dimensions," "obvious fluidity," and occasional "drastic shifts" in "every man's [*sic*] self-interest." Organizers speak of seeking to foster "enlightened" or "thick" as opposed to "narrow" forms of self interest (Chambers, 2018; Phulwani, 2016). In practice, however, self interest in Alinskyite community organizing is often taken as a given, a starting point, a bedrock to be uncovered and made known; and it is unclear exactly how self interest might be learned, unlearned or relearned. In part, this is due to a general aversion to organized forms of formal education and preference for informal learning through direct experience of community organizing. "Self interest is not thickened through reflection, but as part of the activity of organizing itself," notes Phulwani (2016, p. 869; see also Alinsky, 1969, p. 169). But it is not clear how this happens, or what conditions need to be present for "thickening" of self interest to occur effectively; and it begs the question of how people can be persuaded to join an organization taking action in the first place, if they don't yet see climate crisis action as something that benefits their own interests. If we are going to foreground a concern with self interest in climate change education, we need to have a clearer understanding of how self interest might be transformed through education, and not treated as something that either is fixed in perpetuity or changes in mysterious and unknown ways.

Further, although Alinsky (1969, pp. 167, 169) clearly points to the importance of a popular education process that can "shatter the shell of isolationism" surrounding the local community, through analysing and understanding how community interests and concerns are shaped by the "vast conflicts, pressures, stresses and strains of the entire social order," in practice, such learning tends to be underdeveloped in Alinskyite community organizing. Learning in this approach is more likely to be focused on learning about local issues and interests, identifying targets and opponents and their often antithetical interests, and developing effective strategies, tactics and actions for putting pressure on these targets and winning campaign goals: this is the principal strength of this kind of community organizing (Beck & Purcell, 2013; Sen, 2003; Walls, 2015). But broader, foundational forms of political education are often limited. One reason for this is Alinsky's aversion to anything that seems "ideological," and aspiration to develop campaigns that are allegedly "free of ideology" (Sen, 2003, p. xlvi). Alinsky's emphasis on individual and community self interest was intended as an alternative to engaging with ideology, which he saw as being prone to dogmatism and factionalism, and politically and practically ineffectual or harmful (Stall & Stoecker, 1998). This stance, however, risks leaving the Alinskyite approach without the kind of "pedagogy of hope" that (as discussed in Chap. 4) many social justice educators argue is vital for effective learning and organizing for change: the development of what hooks (2003) calls a "prophetic imagination" of a better future world, along with what Freire (1994) describes as a "serious, correct political analysis" of the current "situation of oppression." Critics of Alinsky like Rinku Sen (2003, pp. 7, 21, 166) argue that it is vital that community organizers engage in ideological work, political education and consciousness raising, to understand "how society is organized, why it is organized that way, and how change will come," and "to articulate a set of values" and visions that will inspire and mobilize large numbers of people (see also Beck & Purcell, 2013; and Walls, 2015). This then raises a question about the claim made by Katharine Hayhoe at the outset of this chapter, that we can achieve the radical social transformation needed to address the climate crisis by simply "connecting the dots" between climate action and self interest, without having to try to shift people's values, visions, ideas or ideologies. Self interest is an essential place to begin any effective climate change education: but perhaps it is less helpful if accepted as an end point for that education as well.

Democratic Education and Consciousness Raising

Thus, if effective climate change education needs to begin with the interests of learners—with what matters to them in their own lives—it also needs to recognize that the formation and articulation of self interest, whether in relation to the climate crisis or other matters, is itself a vital site for education. Self interest is something that always needs to be learned, and is not necessarily automatic or self evident to us as individual selves. Though at first this may seem absurd—of course I know what matters in my own life!—on reflection, it becomes clear why this is the case. Erik Olin Wright (2019) points out that individual interests, which he defines as "things that would make a person's life go better along some dimension important to that person" and as "anchored in the solutions to the problems people encounter in their lives," are complex, multiple and often contradictory, as they are linked to an individual's "class location, gender, health status, religion, ethnicity, nationality, language, sexuality." People may have "short-run interests and long-term interests, which may also be in tension." This means, Wright points out, that "when people think about what is in their interests, they inevitably have to foreground some interests and bracket others." It also means that we need to understand how our interests are linked to a larger group: my struggles with unemployment or sense of disconnect with climate change activism, are not simply an individual problem, but based, perhaps, in my race or class location. At different moments in our lives, we may see our interests as being more local, global or universal in scope. To understand our own self interest, then, we need to develop our "sociological imagination," that "enables us to grasp history and biography and the relations between the two in society" (Mills, 2000, p. 6).

Interests are things that we may be more or less consciously aware of and reflexive about. "Most people are too immersed in daily life to be aware of how they constitute their worlds," argues Maxine Greene (1974): "Taking for granted the commonsense appearances of things, governing themselves by the recipes others impose on them …, they function habitually and compliantly" (quoted in Chilcoat & Ligon, 1998, p. 190). We may automatically assume certain interests because of our belonging to a particular social group, when there is actually a need and opportunity to rethink whether these really are the interests of myself or my group. We may not always be aware that our actions, attitudes and professed values may be shaped by our own interests, the stances we take in the world more particular than we like to think. As interests are related to perceived solutions to problems that we face, as "a kind of prediction about the effects of alternative possibilities" or a belief about "the effects of different courses of action," our understanding of our own self interest relies on evidence, theory, argument and, quite often, political ideology (Wright, 2019). It is thus quite possible to be mistaken in our claims of self interest, to be subject to "false consciousness," in the sense of having "a false understanding of what in fact would make one's life better," or "what specifically are the best means to realize some end" (Wright, 2019).

All of this means that the learning, unlearning and relearning of self interest is an essential educational project, whether in relation to the climate crisis or any other issue. The body of practice and theory that has done most to think this through is work done on democratic education. Democracy, after all, can be defined as a society in which all people are able "to participate meaningfully in decisions about things which affect their lives"—the things that matter to them, their interests—where the things that affect our lives include not just our "lives as separate persons" but our "lives as members of a broader community" (Wright, 2010, p. 8). Some (liberal) democratic traditions, similar to Alinskyite community organizing, tend to treat interests or preferences as being relatively fixed and exogenous to the democratic process itself, where individuals come together to negotiate and vote on the basis of their own pre-existing self interests (della Porta, 2013, p. 8; Mansbridge, 2000). Other (deliberative) democratic traditions tend to see self interest as something that needs to be transcended, so that through argument, reason and collective deliberation, self interest is supplanted with a commitment to supporting common, collective and public interests above all else: here, the self interest that is to be decentered is understood to be a narrow, private, materialist, individualistic and base form of self interest (della Porta, 2013, pp. 61–62; see also Fraser, 1997, p. 87; Mansbridge et al., 2010; Young, 1996).

It is feminist political theorists who have made the strongest arguments that learning and relearning of self interest is itself a fundamental component of the democratic process. "One of the principal aims of deliberation … [is] helping participants clarify their interests," argues Nancy Fraser (1997, p. 87). Rather than assuming that the ideal outcome of democratic deliberation must always be "the discovery of a common good in which conflicts of interest evaporate as merely apparent," it is also possible and sometimes desirable that deliberation will lead instead to "the discovery that conflicts of interests are real and the common good is chimerical," an ideological "mystification" that may serve the interests of dominant groups in society above all others (Fraser, 1997, pp. 87–88). Individual preferences and interests "are not given" (Mansbridge, 2000). Rather, as Seyla Benhabib (1996, p. 71) notes, in parallel with the claims made by Erik Olin Wright above:

> On complex social and political issues … individuals may have views and wishes but no ordered set of preferences. … It is actually the deliberative process itself that is likely to produce such an outcome by leading the individual to further critical reflection on his [*sic*] already held views and opinions; it is incoherent to assume that individuals can start a process of public deliberation with a level of conceptual clarity about their choices and preferences that can actually result only from a successful process of deliberation. Likewise, the formation of coherent preferences cannot precede deliberation; it can only succeed it. Very often individuals' wishes as well as views and opinions conflict with one another. In the course of deliberation and the exchange of views with others, individuals become more aware of such conflicts and feel compelled to undertake a coherent ordering. More significantly, the very procedure of articulating a view in public imposes a certain reflexivity on individual preferences and opinions. When

presenting their point of view and position to others, individuals must support them by articulating good reasons in a public context to their co-deliberators.

"Deliberation, and the political process more broadly speaking, ought to make participants more aware of their real interests," writes Jane Mansbridge (2000), by providing a space in which claims about interests "must be tentatively voiced, tested, examined against the causes that produced them, explored, and finally made one's own." Depending on the issue at hand, as well as the stakeholders involved, democratic discussion should have "the dual aim of both searching for consensus and clarifying conflict" (Mansbridge, 2017, p. 110).

How can such development of critically reflexive self interest happen? In educational theory and practice, this is described as a process of conscientization and liberation education (Freire, 2000), education as a practice of freedom (hooks, 1994), consciousness raising (Boler, 1999; Bricker-Jenkins & Hooyman, 1986), the fostering of wide-awakeness (Greene, 1995), and so on. In all of these approaches, there is a central concern with fostering a grounded, critical, political and agentive understanding among learners of their own place in the world, through collective learning about identities, ideas and values—but also, learners' own self interest as members of a broader social group, whether this be defined in terms of race, class, gender or some other category, such as the colonized or oppressed. As Paulo Freire (2000, pp. 83, 85) writes, for example, with a liberatory approach to education, learners can "develop their power to perceive critically *the way they exist* in the world *with which* and *in which* they find themselves," and this "deepened consciousness of their situation" can enable them "to apprehend that situation as an historical reality susceptible of transformation" (italics in original).

More practically, such approaches have tended to embrace an overlapping set of strategies. One is the creation of spaces and institutions within which such a politicized and critical process of learning can take place, in recognition of the fact that formal institutions of schooling do not always provide such opportunities. Nancy Fraser (1997, p. 81) calls these spaces subaltern counterpublics, which she describes as "discursive arenas where members of subordinated social groups invent and circulate counter-discourses, which in turn permit them to formulate oppositional interpretations of their identities, interests, and needs." A second common strategy is a commitment to beginning with the experiences, needs and interests of learners. "The starting point for organizing the program content of education," writes Freire (2000, p. 95), "must be the present, existential, concrete situation, reflecting the aspirations of the people." A third is the embrace of an explicitly political curriculum that "promotes on the part of students an informed and critical political and historical consciousness": here, "contentious social and political issues" are "sought out, emphasized, and interrogated," as students are "taught to be suspicious of the status quo and skeptical about the claims of the powerful," and "made sensitive to undemocratic and antidemocratic forces and to the forms and sources of injustice and oppression" (Sabia, 2012, p.389). A fourth strategy is the fostering of open-ended dialogue and deliberation around both local experience and an overtly political curriculum, where dissent and difference are welcome, and learners are given space to make their own decisions about the ideas, values and interests that matter most to them. A fifth is directly linking such spaces of critical reflection to collective social action because, as with Alinskyite community organizing, action is seen to be an essential way to learn about interests, ideas and values in society, but also because all of these approaches seek to use education to support projects of social transformation.

Nancy Fraser (1997, p. 81) thus points to the example of "the late-twentieth-century US feminist subaltern counterpublic," which created an "array of journals, bookstores, publishing companies, film and video distribution networks, lecture series, research centers, academic programs, conferences, conventions, festivals, and local meeting places." Through these spaces, feminist women "recast [their] needs and identities," by developing "new terms for describing social reality, including 'sexism,' 'the double shift,' 'sexual harassment,' and 'marital, date, and acquaintance rape.'" Anita Krajnc (2000, p. 346) looks to the early twentieth century US labor movement's broad-based educational efforts to raise "worker class consciousness," which they recognized as being "far from automatic," through developing "their own press, mass meetings, pickets, labor schools, songs [and] theatre." George Chilcoat and Jerry Ligon (Chilcoat & Ligon, 1998, pp. 171, 173) describe how the Freedom

Schools, run by the US civil rights movement in Mississippi in the early 1960s, used a combination of a political curriculum, open-ended discussion and direct social action to "raise black students' consciousness" and enable them "to become a force for social change":

> A liberating, democratic education for Mississippi black students was the objective of freedom schools. … Students would be encouraged to ask questions about their experiences and their personal situations. … Asking questions was to be the means for converting attitudes from passive to active—helping students 'to understand themselves, and thereby to understand their society and the need to change it. … The idea is to teach Negro Mississippians to take themselves seriously, to articulate their ambitions, and their discontents—in short, to instill political awareness. (pp. 172–173)

In all of these and other similar examples of democratic, political education, a key goal is not just learning to "connect the dots" of how wider social issues—like the climate crisis—affect one's self interests. It is also to develop and often radically transform what one understands one's self interests (and identity and values and ideas) to be in the first place. The reason this is vital is that often, initial understanding of self interest may be shaped and constrained by dominant discourses of patriarchy, capitalism, racism, colonialism, extractivism and so on—the very same ideologies that are at root of broader social problems (including the climate crisis) to begin with. In such cases, it is never going to be sufficient to link social problems to self interests, but will require a critically reflexive rethinking of these self interests as well.

Critical Pedagogy and the Contestation of Dominant Group Interests

One reason why educational projects to foster critically reflexive understanding of the self interests of disempowered groups in society have often taken place outside of the school system, is the recognition that formal education, in many contexts, is organized in support of dominant group interests. Paulo Freire's *Pedagogy of the Oppressed* represents one of the clearest and most influential articulations of this recognition. For Freire, all education and learning is inherently political, in that it is always shaped, constrained and oriented to some set of interests, values and ideas over others. At some level, writes Freire (2000, p. 81), education works as either a "practice of domination," and/or a "practice of freedom." "There is no truly neutral education," Freire (1985) argues: "'Washing one's hands' of the conflict between the powerful and the powerless means to side with the powerful, not to be neutral." Schooling—"systematic education" is Freire's term—is often shaped by the interests, ideas and values of dominant groups and actors. "Dominant elites," claims Freire (2000, pp. 78, 95, 147), seek to use formal education with an "ideological intent" of "indoctrinating [students] to adapt to the world of oppression," to "encourage passivity," and "conform the masses to their objectives." A central objective of liberation education (or a "pedagogy of the oppressed"), therefore, is not just developing a deeper self consciousness of the interests and needs of marginalized groups, but also a clear analysis of the interests, ideas and values of dominant groups, and articulation of a rationale and strategy for why and how these groups' agendas should be contested and displaced. In some contexts, it might be deemed possible to do this from within the school system itself; at other times, it might be decided that it is politically more effective to create alternative educational spaces (Fraser's subaltern counterpublics) altogether.

To develop an educational approach that can effectively address the challenge of the climate crisis, or any other social or environmental problem, a central task is to identify, analyse and contest the interests of dominant groups and actors who may stand in the way of effective climate change education and action, to develop a critical understanding of climate change politics *and* climate change education politics. One key reason why climate action has been so limited over the past four decades, since leaping into widespread public, media and political consciousness in the 1980s, is not, contrary to the claims of some, due to failings of "human nature," such as an allegedly innate incapability of "sacrificing present convenience to forestall a penalty imposed on future generations": rather, it is due in large part to systematic opposition by elites, who have used this period to further their own interests by promoting a neoliberal agenda of extending "privatization, deregulation, and economic austerity to every corner of the globe" (Klein, 2019, pp. 245, 247, 248). This analytic task is one that is embraced by the tradition of critical pedagogy, since a central question critical pedagogy always asks is: "Whose interests are being served and at what price?" (White et al., 2014, p. 133). "The critical question," as Henry Giroux (2007) says about critical pedagogy, "is whose future, story and interest does the public school, higher education, or any other educational site represent." Whose interests are served by a particular curriculum, pedagogy or education policy? Whose interests are served by current ways of organizing our society and economy?

When considering the climate crisis, this question of interests is vital to ask within the field of education. Emily Eaton and Nick Day (2020, pp. 457, 470) have coined the term petro-pedagogy to refer to the ways in which "fossil fuel interests reach into schools" to promote "teaching practices and resources" that are "designed to insulate fossil fuel industries from criticism and dissuade young people from questioning or understanding the role of corporate power in the climate crisis." Eaton and Day argue that "environmental education must begin to teach about the power, influence, and interests of fossil capital" (p. 14). In previous work, I used Eaton and Day's concept of petro-pedagogy to examine how fossil fuel corporations have embedded themselves "at the heart of elite … science and education policy and practice networks," notably "networks focused on the development of STEM (Science, Technology, Engineering and Mathematics) education," where they have become core participants in "a more extensive corporate education reform network that, for the past decade, has focused on promoting a neoliberal model of STEM education in schools around the world, based firmly on a 'corporate/ capitalist vision for the purposes of schooling.'" (Tannock, 2020, p. 2). More generally, researchers such as Wayne Au and Joseph Ferrare (Au & Ferrare, 2015) have investigated how corporate education reform networks have actively promoted neoliberal models of public education that serve their own political and economic interests. These neoliberal models (discussed in Chap. 5) impede the ability of schools, teachers and students to develop effective programs of climate change education and action, through marginalizing the concerns of environmental and climate change education in the school curriculum, and working to depoliticize and individualize this education when it does appear.

It is also vital to ask questions of the role of elite interests in blocking, undermining or redirecting effective education and action on the climate crisis more generally, beyond the formal education system. Here, extensive research and analysis has been done on how "fossil fuel interests have continued to spend tens of millions of dollars in an online public-relations campaign to discredit the science of human-caused climate change" (Mann & Toles, 2016, p. 34). But it is not just the fossil fuel sector that needs to be considered. Corporate business groups, across sectors, have sought to guide climate change policy in line with their interests, by insisting that "the answer to climate change should inevitably revolve around market expansion and economic growth" and acting "to maintain the hegemony of neoliberal capitalism" (Wright & Nyberg, 2015, pp. 73–74). "Commercial interests and influence of corporate lobbies," argue Sula-Raxhimi et al. (2019, p. e12), are "the biggest challenge to governments' lack of engagement" with climate change. Climate apartheid, in which "the wealthy pay to escape overheating, hunger, and conflict, while the rest of the world is left to suffer," has emerged as a growing threat to climate change education and action seeking to address the root causes of climate change, while ensuring that the disadvantaged are as protected from the impacts of climate crisis as the privileged (García, 2020). In today's climate change politics, writes Tom Whyman (2019), "the selfish interests of the rich already diverge from those of the poor:"

> Last year's record-breaking wildfires in California were fought, in part, by private firefighters hired by insurance companies to protect only the houses of those able to afford their services. … In Miami, rising sea levels are driving gentrification, with richer residents buying houses in traditionally poorer areas that sit on higher ground; those they displace are often forced to seek affordable housing in areas where flooding is more common.

The super-rich are buying "luxury bunkers," "apocalypse holiday homes," and properties in "locations safe from the effects of climate change," not as part of a combined effort to save all of humanity from the effects of the climate crisis, but to ensure the security and well-being of themselves, their families and loved ones only (Cooke, 2019; Osnos, 2017).

In interrogating dominant group interests and agendas, a key principle of critical pedagogy is to embrace a structural and systemic analysis. The aim is not just to identify and analyse dominant actors and their interests, but the ideological, social and economic systems that shape and constrain actors and their interests, from dominant and marginalized groups alike. Freire (2000, pp. 58, 81, 154) describes this as a "constant unveiling of reality," through the identification of "situations of oppression" and their underlying "structures of domination." By now, there is a rich and continually growing literature from sociology, anthropology, geography, history and political economy for a critical pedagogy of the climate crisis to draw upon, that offers careful and detailed analysis of the role of capitalism, colonialism and imperialism, systemic racism, patriarchy, and other social, economic and technological systems in producing the climate crisis and blocking effective projects of education and action to address the crisis (see, for example, Angus, 2016; Buckingham & Le Masson, 2017; Godfrey & Torres, 2016; Hickel, 2020; Holleman, 2018; Mahony & Endfield, 2018; Pirani, 2018; Yusoff, 2018).

In all of these, emphasis is placed on linking actors, actions and interests to broader social, economic, technological and ideological systems. Climate apartheid, argues Nancy Tuana (2019, pp. 5–6), is shaped not simply by the actions of privileged groups, but "deeply held systematic beliefs and dispositions regarding racial superiority" that shape "individual beliefs and practices" and "social institutions," and intersect with "other forms of systemic oppression such as those due to gender, sexuality, or class. "The relentless expansionary drive of corporations" that is responsible, in part, for growing climate and environmental crises, is not driven solely by CEO obsession "with accumulating money and power," writes Jason Hickel (2020, p. 86): "The reality is that these firms, and the CEOs who run them, are subject to a *structural imperative* for growth" that defines capitalism as a social and economic system (italics in original). "Capital doesn't *mean to* destroy … wild nature," claims Andreas Malm (2020, p. 59), in the sense that "it doesn't have an intention formed in the mind and then engage in efforts to realise it—there is just no other way for it to replicate" (italics in original). This is the sense in which

Freire (2000, pp. 56, 57) claims, in *Pedagogy of the Oppressed*, that "an oppressive class, can free neither others nor themselves," for they themselves are "submerged" in a "situation of violence and oppression" and "caught up in … an entire way of life and behavior." This task of freedom, transformation—and climate justice—is one that falls to a project of critical pedagogy, conducted by those who have the greatest interest and need to take urgent action for radical social change.

Instrumental and Emancipatory Models of Climate Change Education

There is currently limited mention of democratic education or critical pedagogy in discussions of climate change education. To return to the suggestions made by Katharine Hayhoe at the start of this chapter, Martha Monroe and her colleagues found in their systematic review of climate change education research that only a "very few articles" embraced pedagogical approaches that had the goal of enabling learners "to think about what really and profoundly matters, to collectively envision a better future, and then to become practical visionaries in realizing that future" (Monroe et al., 2019, p. 807). Busch, Henderson and Stevenson (2019, p. 962), likewise, note that "few articles" on climate change education research start "from a point of normative contestation, whereby dominant political, economic, and sociocultural trends are interrogated in order to change them via educational praxis," just as few "focus attention to the politics and policies that hinder … or promote … climate change education." Where we find consideration of these educational approaches is in the work of educators, researchers and theorists who argue for the importance of differentiating between instrumental and emancipatory models of environmental education.

Arjen Wals and Justin Dillon (2012, p. 255) argue that environmental education has been dominated by instrumental forms of education that aim "at changing learner behavior" in relation to the environment—where "behavior" includes learners' "awareness, attitudes, beliefs, and values"—in order "to achieve predetermined goals" of addressing environmental destruction and crisis. The problem with instrumental approaches, as critics see it, is partly that the focus on individual behavior change is often ineffective. As discussed in Chap. 3, not only do such models often fail to transform individual actions in the way they hope, they also fail to address the need to transform social, economic and ideological structures and institutions, to effectively address environmental and climate crisis. Further, some critics of instrumental approaches "argue that using education to change peoples' behavior in a pre- and expert-determined direction has more to do with manipulation and indoctrination than with education" (Wals et al., 2008, p. 56).

In contrast, critics argue for an emancipatory approach to environmental education that focuses on "learning, capacity building, and critical thinking that will allow citizens to understand what is going on in society, to ask critical questions, and to determine for themselves what needs to be done" (Wals & Dillon, 2012, p. 256). Here, educators focus on fostering student skills in "anticipatory thinking, integrative thinking, dealing with complexity and ambiguity, [and] action competence," to strengthen student "capacities for critical engagement in the key issues of our time" (Peters & Wals, 2016, p. 183). But the dilemma with this emancipatory approach, these critics worry, is that students may decide not to focus urgently on issues like the climate crisis, that is, after all, a highly time sensitive matter; indeed, they may not prioritize sustainability as a core value and interest at all (Wals & Jickling, 2002). It seems there might be a "tension," Wals (2010, p. 150) suggests, between democratic education's "call for self-determination on the one hand, and the call for sustainable social norms and education for sustainable development as a destination on the other."

The discussion in this chapter of the role of self interest in community organizing, democratic education and critical pedagogy may help resolve some of this apparent tension between embracing emancipatory models of education and fostering forms of education that can effectively address the climate crisis. One problem in opposing instrumental and emancipatory models of education in this way is it risks misrepresenting the relationship of directiveness to democratic, liberation or emancipatory education, where directiveness concerns the role of educators in "stimulating learners, pointing to objectives, and aiming to convince learners of some knowledge or belief," such as the nature and significance of the climate crisis (Chambers, 2019, p. 23). Freire (1994, p. 65) insists that "education of its very nature is directive and political." "Whether it be authoritarian or democratic," Freire argues, "educational practice … is always directive" (p. 66).

But another problem is understanding the importance of interest formation and transformation as a core component of democratic education. A defining characteristic of democratic education is that, unlike instrumental models of education, it always starts with the interests of learners: it doesn't ignore these or seek to impose unilateral educational imperatives from above. But this is not to suggest that democratic education does not, or should not, seek to develop and often radically transform the interests (or, to use the words of Wals and Dillon, the "awareness, attitudes, beliefs, and values") of students. On the contrary, this is one of the most essential things democratic education can do. All of the examples of democratic, liberation and critical education discussed in this chapter seek to do this. The key question is why and how this is done: critical, democratic educators have, as noted above, commonly embraced strategies such as the creation of alternative spaces of learning, development of an explicitly political curriculum, embrace of open-ended student dialogue, invitation to participate in collective action, and so forth. Through these and other such approaches, a democratic climate change education would take student self interest as a primary concern and starting point for the work it seeks to do; but also as a focus for directive, dialogical, collective and critical reflection, learning and change.

References

Alinsky, S. (1969). *Reveille for radicals*. Random House.

Alinsky, S. (1971). *Rules for radicals: A pragmatic primer for realistic radicals*. Vintage Books.

Angus, I. (2016). *Facing the Anthropocene: Fossil capitalism and the crisis of the earth system*. Monthly Review Press.

Aronoff, K., Battistoni, A., Cohen, D., & Riofrancos, T. (2019). *A planet to win: Why we need a Green New Deal*. Verso.

Au, W., & Ferrare, J. (Eds.). (2015). *Mapping corporate education reform.* Routledge.

Ayling, J., & Gunningham, N. (2017). Non-state governance and climate policy: The fossil fuel divestment movement. *Climate Policy, 172*, 131–149.

Beck, D., & Purcell, R. (2013). *International community organising: Taking power, making change.* Policy Press.

Benhabib, S. (1996). Toward a deliberative model of democratic legitimacy. In S. Benhabib (Ed.), *Democracy and difference* (pp. 67–94). Princeton University Press.

Boler, M. (1999). *Feeling power: Emotions and education.* Routledge.

Boström, M., et al. (2018). Conditions for transformative learning for sustainable development: A theoretical review and approach. *Sustainability, 10*(4479), 1–21.

Bricker-Jenkins, M., & Hooyman, N. (1986). Feminist pedagogy in education for social change. *Feminist Teacher, 2*(2), 36–42.

Buckingham, S., & Le Masson, V. (Eds.). (2017). *Understanding climate change through gender relations.* Routledge.

Busch, K., Henderson, J., & Stevenson, K. (2019). Broadening epistemologies and methodologies in climate change education research. *Environmental Education Research, 25*(6), 955–971.

Chambers, E. (2018). *Roots for radicals: Organizing for power, action and justice.* Bloomsbury.

Chambers, D. (2019). Is Freire incoherent? Reconciling directiveness and dialogue in Freirean pedagogy. *Journal of Philosophy of Education, 53*(1), 21–47.

Cheon, A., & Urpelainen, J. (2018). *Activism and the fossil fuel industry.* Routledge.

Chilcoat, G., & Ligon, J. (1998). 'We talk here. This is a school for talking': Participatory democracy from the classroom out into the community: How discussion was used in the Mississippi Freedom Schools. *Curriculum Inquiry, 28*(2), 165–193.

Cooke, E. (2019). Apocalypse holiday homes—How the super-rich are planning for the end of the world. *Telegraph*, 17 October. https://www.telegraph.co.uk/travel/comment/apocalypse-holiday-homes-new-zealand-super-rich/

della Porta, D. (2013). *Can democracy be saved?* Polity Press.

Eaton, E., & Day, N. (2020). Petro-pedagogy: Fossil fuel interests and the obstruction of climate justice in public education. *Environmental Education Research, 6*(4), 457–473.

Fraser, N. (1997). *Justice interruptus: Critical reflections on the 'postsocialist' condition*. Routledge.
Freire, P. (1985). *The politics of education*. Bergin & Garvey.
Freire, P. (1994). *Pedagogy of hope*. Bloomsbury.
Freire, P. (2000). *Pedagogy of the oppressed*. Continuum.
Friedman, L. (2017). Hurricane Irma linked to climate change? For some, a very 'insensitive' question. *New York Times*, 11 September. https://www.nytimes.com/2017/09/11/climate/hurricane-irma-climate-change.html
García, M. (2020). The media isn't ready to cover climate apartheid. *Nation*, 17 June. https://www.thenation.com/article/environment/climate-apartheid-inequality-media/
Ghosh, A. (2016). *The great derangement: Climate change and the unthinkable*. University of Chicago Press.
Giroux, H. (2007). Higher education in a time of crisis: Rethinking the politics and possibilities of critical pedagogy. *Cosmopolis*, 7 February. http://www.cosmopolisonline.it/articolo.php?numero=II22007&id=14
Godfrey, P., & Torres, D. (Eds.). (2016). *Systemic crises of global climate change: Intersections of race, class and gender*. Routledge.
Greene, M. (1974). Cognition, consciousness and curriculum. In W. Pinar (Ed.), *Heightened consciousness, cultural revolution and curriculum theory*. McCutchan.
Greene, M. (1995). *Releasing the imagination: Essays on education, the arts, and social change*. Jossey-Bass.
Gunningham, N. (2017). Building norms from the grassroots up: Divestment, expressive politics and climate change. *Law & Policy, 394*, 372–392.
Håkansson, M., & Östman, L. (2019). The political dimension in ESE: The construction of a political moment model for analysing bodily anchored political emotions in teaching and learning of the political dimension. *Environmental Education Research, 25*(4), 585–600.
Hart, S. (2001). *Cultural dilemmas of progressive politics: Styles of engagement among grassroots activists*. University of Chicago Press.
Hayhoe, K. (2016). Why climate change should matter to you. *EcoWatch*, 26 December. https://www.ecowatch.com/hayhoe-climate-change-2117925528.html
Hayhoe, K. (2017). Yeah, the weather has been weird. Foreign Policy, 31 May. https://foreignpolicy.com/2017/05/31/everyone-believes-in-global-warming-they-just-dont-realize-it/

Hayhoe, K. (2019). How to talk about climate change so people will listen. *Chatelaine*, 18 April. https://www.chatelaine.com/living/how-to-talk-about-climate-change/

Healy, N., & Barry, J. (2017). Politicizing energy justice and energy system transitions: Fossil fuel divestment and a 'just transition'. *Energy Policy, 108*, 451–459.

Healy, N., & Debski, J. (2017). Fossil fuel divestment: Implications for the future of sustainability discourse and action within higher education. *Local Environment, 226*, 699–724.

Henderson, J. (2019). Learning to teach climate change as if power matters. *Environmental Education Research, 25*(6), 987–990.

Henderson, J., & Drewes, A. (2020). Teaching climate change in the United States. In J. Henderson & A. Drewes (Eds.), *Teaching climate change in the United States*. Routledge.

Henderson, J., & Zarger, R. (2017). Toward political ecologies of environmental education. *Journal of Environmental Education., 48*(4), 285–289.

Hickel, J. (2020). *Less is more: How degrowth will save the world.* William Heinemann.

Holleman, H. (2018). *Dust bowls of empire: Imperialism, environmental politics, and the injustice of 'green' capitalism*. Yale University Press.

hooks, b. (1994). *Teaching to transgress: Education as the practice of freedom*. Routledge.

hooks, b. (2003). *Teaching community: A pedagogy of hope*. Routledge.

Huber, M. (2019). Ecological politics for the working class. *Catalysts, 3*(1). https://catalyst-journal.com/vol3/no1/ecological-politics-for-the-working-class

Jorgenson, S., Stephens, J., & White, B. (2019). Environmental education in transition: A critical review of recent research on climate change and energy education. *Journal of Environmental Education, 50*(3), 160–171.

Kenis, A. (2019). Post-politics contested: Why multiple voices on climate change do not equal politicisation. *EPC: Politics and Space, 37*(5), 831–848.

Kenis, A., & Lievens, M. (2014). Searching for 'the political' in environmental politics. *Environmental Politics, 23*(4), 531–548.

Kenis, A., & Mathijs, E. (2012). Beyond individual behaviour change: The role of power, knowledge and strategy in tackling climate change. *Environmental Education Research, 18*(1), 45–65.

Klein, N. (2019). *On fire: The burning case for a Green New Deal.* Penguin.

Kollmuss, A., & Agyeman, J. (2002). Mind the gap: Why do people act environmentally and what are the barriers to pro-environmental behavior? *Environmental Education Research, 8*(2), 239–260.
Krajnc, A. (2000). Popular environmental education: Lessons from the labor and civil rights movements. *New Political Science, 223*, 341–360.
MacGregor, S. (2014). Only resist: Feminist ecological citizenship and the post-politics of climate change. *Hypatia, 29*(3), 617–632.
Mahony, M., & Endfield, G. (2018). Climate and colonialism. *WIREs Climate Change, 9*(2), e510.
Malm, A. (2020). *Corona, climate, chronic emergency*. Verso.
Malm, A., & Warlenius, R. (2019). The grand theft of the atmosphere: Sketches for a theory of climate injustice in the Anthropocene. In K. Bhavnani et al. (Eds.), *Climate futures: Re-imagining global climate justice*. Zed Books.
Mangat, R., Dalby, S., & Paterson, M. (2018). Divestment discourse: War, justice, morality and money. *Environmental Politics, 27*(2), 187–208.
Maniates, M. (2001). Individualization: Plant a tree, buy a bike, save the world? *Global Environmental Politics, 1*(3), 31–52.
Maniates, M. (2016). Make way for hope: A contrarian view. In S. Nicholson & S. Jinnah (Eds.), *New earth politics: Essays from the Anthropocene* (pp. 135–154). MIT Press.
Mann, M., & Toles, T. (2016). *The madhouse effect*. Columbia University Press.
Mansbridge, J. (2000). Feminism and democracy. *American Prospect*, 5 December. https://prospect.org/civil-rights/feminism-democracy/
Mansbridge, J. (2017). The long life of Nancy Fraser's 'Rethinking the public sphere'. In B. Bargu & C. Bottici (Eds.), *Feminism, capitalism and critique* (pp. 101–118). Palgrave Macmillan.
Mansbridge, J., et al. (2010). The place of self-interest and the role of power in deliberative democracy. *Journal of Political Philosophy, 18*(1), 64–100.
Marshall, G. (2014). *Don't even think about it: Why our brains are wired to ignore climate change*. Bloomsbury.
Mills, C. (2000). *The sociological imagination*. Oxford University Press.
Monroe, M., et al. (2019). Identifying effective climate change education strategies: A systematic review of the research. *Environmental Education Research, 25*(6), 791–812.
Nilsen, E. (2019). The new face of climate activism is young, angry—And effective. *Vox*, 17 September. https://www.vox.com/the-highlight/2019/9/10/20847401/sunrise-movement-climate-change-activist-millennials-global-warming

Osnos, E. (2017). Doomsday prep for the super-rich. *New Yorker*, 23 January. https://www.newyorker.com/magazine/2017/01/30/doomsday-prep-for-the-super-rich

Peters, M., & Wals, A. (2016). Transgressive learning in times of global systemic dysfunction: Interview with Arjen Wals. *Open Review of Educational Research, 3*(1), 179–189.

Phulwani, V. (2016). The poor man's Machiavelli: Saul Alinsky and the morality of power. *American Political Science Review, 110*(4), 863–875.

Pirani, S. (2018). *Burning up*. Pluto.

Sabia, D. (2012). Democratic/utopian education. *Utopian Studies, 23*(2), 374–405.

Schutz, A., & Sandy, M. (2011). *Collective action for social change: An introduction to community organizing*. Palgrave Macmillan.

Sen, R. (2003). *Stir it up: Lessons in community organizing and advocacy*. Jossey-Bass.

Stall, S., & Stoecker, R. (1998). Community organizing or organizing community? Gender and the crafts of empowerment. *Gender & Society, 12*(6), 729–756.

Stevenson, R. (2007). Schooling and environmental education: Contradictions in purpose and practice. *Environmental Education Research, 13*(2), 139–153.

Sula-Raxhimi, E., Butzbach, C., & Brousselle, A. (2019). Planetary health: Countering commercial and corporate power. *Lancet, 3*, e12–e13.

Sund, L., & Öhman, J. (2014). On the need to repoliticise environmental and sustainability education: Rethinking the postpolitical consensus. *Environmental Education Research, 20*(5), 639–659.

Swyngedouw, E. (2010). Apocalypse forever? Post-political populism and the spectre of climate change. *Theory, Culture & Society, 27*(2), 213–232.

Tannock, S. (2020). The oil industry in our schools: From Petro Pete to science capital in the age of climate crisis. *Environmental Education Research, 26*(4), 474–490.

Tuana, N. (2019). Climate apartheid: The forgetting of race in the Anthropocene. *Critical Philosophy of Race, 7*(1), 1–31.

Van Poeck, K., & Östman, L. (2018). Creating space for 'the political' in environmental and sustainability education practice: A political move analysis of educators' actions. *Environmental Education Research, 24*(9), 1406–1423.

Walls, D. (2015). *Community organizing: Fanning the flame of democracy*. Polity Press.

Wals, A. (2010). Between knowing what is right and knowing that it is wrong to tell others what is right: On relativism, uncertainty and democracy in environmental and sustainability education. *Environmental Education Research, 16*(1), 143–151s.

Wals, A., & Dillon, J. (2012). Conventional and emerging learning theories. In R. Stevenson et al. (Eds.), *International handbook of research on environmental education* (pp. 253–261). Routledge.

Wals, A., & Jickling, B. (2002). 'Sustainability' in higher education from doublething and newspeak to critical thinking and meaningful learning. *Higher Education Policy, 15*, 121–131.

Wals, A., et al. (2008). All mixed up? Instrumental and emancipatory learning toward a more sustainable world. *Applied Environmental Education and Communication, 7*(3), 55–65.

Watts, J. (2019). A thermometer is not liberal or conservative. *Guardian*, 6 January. https://www.theguardian.com/science/2019/jan/06/katharine-hayhoe-interview-climate-change-scientist-crisis-hope

White, R., Cooper, K., & Mackey, W. (2014). Culturally relevant education and critical pedagogy. *Revista Internacional de Educación para la Justicia Social, 3*(2), 123–140.

Whyman, T. (2019). Will the rich escape climate apocalypse? *New Internationalist*, 17 May. https://newint.org/features/2019/04/09/first-class-lifeboats

Wright, E. (2010). *Envisioning real utopias*. Verso.

Wright, E. (2019). *How to be an anticapitalist in the twenty-first century*. Verso.

Wright, C., & Nyberg, D. (2015). *Climate change, capitalism and corporations*. Cambridge University Press.

Young, I. (1996). Communication and the other: Beyond deliberative democracy. In S. Benhabib (Ed.), *Democracy and difference* (pp. 120–136). Princeton University Press.

Yusoff, K. (2018). *A billion black Anthropocenes or none*. University of Minnesota Press.

8

Learning Power & Taking Collective Action

On the internet, there is a short video that was produced by a group of secondary students from Notre Dame Girls' School in South London, a school made up almost entirely of black and minority ethnic students, many from low income and immigrant families. In the video, a team of nine students introduce and role play the key steps they say one needs to take to make progressive change in society:

> *Step One*: First, if you want to make change, you have to have power.
> *Step Two*: If you want power, you need people.
> *Step Three*: We then listen to people within our communities to find out what the issues are.
> *Step Four*: Then we take action. We usually take action to get a meeting with the decision maker around the issue we have chosen to try to change.
> *Step Five*: We then negotiate with decision makers and politicians.
> *Step Six*: We then evaluate to see what we can improve and think about our next steps.
> (Notre Dame Girls' School, 2019)

S. Tannock, *Educating for Radical Social Transformation in the Climate Crisis*, Palgrave Studies in Education and the Environment,
https://doi.org/10.1007/978-3-030-83000-7_8

The steps to social change presented by the Notre Dame students are a version of what is more widely known as the cycle of community organizing (McKibban & Steltenpohl, 2019; Speer et al., 1995; Walls, 2015). The students learned the steps through participating in their school's partnership with South London Citizens, a chapter of the nationwide Citizens UK community organizing alliance. This is a partnership that has seen students at Notre Dame push for a range of changes in their community over the past few years: from challenging the UK government over the high cost of child citizenship fees that are preventing children from registering as British citizens; to seeking to change the narrative around young people in London, by asking local police to stop promoting negative and harmful stereotypes of young people, and focus on improving often conflictual relationships between the police and local youth (Arkwright, 2020).

Though short, the Notre Dame Girls' School student video focuses on something which has been missing from most discussions of climate change education, and yet which is central to any possibility of effectively addressing the climate crisis: the importance of power and taking collective action. These are the first steps the students point to in learning how to make changes in society, arrived at within seconds of the opening of their video. In climate change education, on the other hand, one often has to look long and hard before being able to find any similar discussions. As Joseph Henderson and Andrea Drewes (2020, p. 5) write, in their edited collection on climate change education in the United States:

> While climate change education has a decent history of teaching the physical mechanisms of climate change, it lacks much of an analysis of social or political power and is often unwilling to engage those crucial aspects of life.

Others have made parallel observations. Magnus Boström et al. (2018, p. 4479) argue that, "even if there are exceptions, within the greater body of literature on learning for sustainable development, … power remains largely undiscussed and untheorized." Simon Jorgenson, Jennie Stephens and Beth White (Jorgenson et al., 2019, p. 166), in a review of climate change education research, similarly worry that "environmental educators and researchers are neglecting the politics and policy of energy system transformation" and "minimizing the role of collective action." As discussed in Chap. 2, much climate change education focuses narrowly on enhancing climate science literacy; while more ambitious programmes are most likely to focus on facilitating shifts in knowledge paradigms. When action is invoked in climate change education, it is often individualized, apolitical and non-confrontational. Discussions of power, too, tend to be found in relation to agendas of individual "empowerment," that look inward and refer to changes in individual attitudes, abilities and behaviors brought about by increased knowledge and understanding of the climate crisis. Thus, students may be "empowered" by climate change education to adopt sustainable lifestyles and consumer choices in the context of their own homes, families and daily routines.

At the same time, there is widespread recognition that questions of power and collective action are central to addressing the climate crisis. This crisis, as Naomi Klein (2014) writes, "has a lot less to do with the mechanics of solar power than the politics of human power—specifically whether there can be a shift in who wields it, a shift away from corporations and toward communities." "Actually shifting the social and material conditions of climate change," argue Henderson and Drewes (2020, p. 5), "means confronting entrenched systems of power." "If learning activities do not engage in issues of structural inertia, power, inequality, vested economic interests, denialism, [and] resistance to change," warn Boström et al. (2018, p. 4479), "that is, in the *conflict dimension* of sustainable development, … these will likely be insufficient in terms of transforming society in a sustainable direction." In this chapter, I look at the rise of power analysis in the climate change movement, before turning to consider the traditions of social movement learning, community organizing and popular education, on which climate change activist groups who are interested in learning how to analyse and build social power tend to draw. In the final section of the chapter, I consider whether and how this question of learning power and collective action could ever be fully embraced by climate change education as practiced in schools, universities and the rest of the formal education sector.

Learning Power in the Climate Movement

Where we can find more discussion of learning how to analyse, contest, build and shift social power is away from the literature on climate change education, and away from formal education classrooms, in activist groups participating in the climate movement. Over the last couple of decades, there has been a marked turn within the climate movement to focus on the question of how to build and mobilize movement power to the degree that it can overcome the power of vested interests—such as the fossil fuel industry—in preserving the status quo, and begin pushing through the kinds of radical social, political and economic changes needed to address the climate crisis. This turn to a concern with power can be seen in at least three distinct if overlapping sectors of the movement: local, often Indigenous led campaigns to stop proposed fossil fuel extraction and pipeline projects; youth and student led climate action groups; and mass civil disobedience events, most notably, those planned and practiced by the Extinction Rebellion network.

The first of these is what Klein (2014) has dubbed "Blockadia," that she describes as a "grassroots pro-democracy movement" and "roving transnational conflict zone that is cropping up with increasing frequency and intensity wherever extractive projects are attempting to dig and drill, whether for open-pit mines, or gas fracking, or tar sands oil pipelines." This includes high profile campaigns against the Keystone XL and Dakota Access pipeline projects in the United States, Tar Sands development and Trans Mountain pipeline in Canada, Ende Gelände coal mine protests in Germany, as well as hundreds of other similar campaigns throughout the world (D'Arcy et al., 2014; Martinez-Alier, 2021; Temper et al., 2020). While many of these campaigns foster the same kinds of climate change and social justice education that have been discussed throughout this book (for example, see Eagle Shield et al., 2020), they have also developed forms of learning specifically around issues of power and collective action. Climate activists in these campaigns learn the relative power and effectiveness of different kinds of "direct action, legal action, or electoral and legislative action" (LeQuesne, 2019, p. 19). They learn to use "confrontational techniques" such as "mass mobilizations and public demonstrations, protest actions, trespass and breaking in, sit-ins and office takeovers, symbolic use of media, blocking roads, eco-sabotage and riots," as well as "conciliatory or conventional techniques" including "behind the scenes negotiation or cooperation, petitions, leafleting, website, information table, lobbying of politicians, media advertising and press releases" (Bradshaw, 2015, p. 437). They learn the importance of discursive strategies to build public support for their campaigns and undermine the legitimacy of the fossil fuel industry, by "fighting the battle of the story" and creating "narratives we use to make meaning of the world around us and shape common sense" (LeQuesne, 2019, p. 20). Campaign activists learn about "the issues, actors and relations of power" involved in fossil fuel development projects, such as:

> which actors benefit and lose in the construction of oil pipelines and how; what vested interests large oil, construction, and shipping corporations have, how different levels of government, politicians, communities, and groups are invested in or opposed to a fossil fuel project; what legal rights local people have to make decisions about pipeline construction; and what rights and power are denied to them. (Kluttz & Walter, 2018, p. 101)

They also often have to learn about how to protect themselves, during their campaign, from attack and removal by private security and police. Kelly Hayes (2020, p. 20), a community educator who ran workshops during the Dakota Access Pipeline protests at Standing Rock, talks of how she ran self-protection and defence lessons for child and adult protesters alike:

> Visitors and long-term residents of the [protest] camps learned how to treat [protesters] harmed by tear gas and pepper spray. Some even learned how to build makeshift gas masks. I reviewed blockade tactics with [protesters] of all ages, but when a nine-year-old watched me put together a blockade device called a lockbox and asked if he could help, I looked to his mother, certain she would say no. She looked down at me and my power drill and said, 'He needs to know.' … The oil industry and the government were attacking and effectively waging war against families, against community, against sovereignty, and against the Earth. It made sense to me, in that

moment, that resistance was becoming not only a community value but a family value.

A second, connected sector of the climate movement that has fostered a growing concern with analyzing and building social power is the student and youth led part of the movement. In 2007, Bill McKibben and six university students in the United States created 350.org, based on an explicit theory of power and social change: corporations and states will only take substantial action to address the climate crisis if they are pressured by a broad based, grassroots movement that can mobilize significant people power (Gunningham, 2019; Hestres & Hopke, 2020). "Our fight against climate change is a battle over power," a 350.org (2019) training video explains, "but not only the kind of power that runs our cars or keeps our buildings warm." More specifically, 350.org argues that climate activists need to directly challenge the economic and political power the fossil fuel industry holds in contemporary society, by dismantling their "social license"; they need to push governments to take urgent climate change action; and they should do this through mass mobilizations, "iconic battles" against the fossil fuel industry, "locally appropriate campaigns" guided by "communities impacted by the exploitation of fossil fuels," and "creative, open tactics that vary across the world" (350.org, 2021; see also Hestres, 2015). 350.org was involved in the the Keystone XL pipeline campaign and created the fossil fuel divestment campaign discussed in Chap. 7. Other youth climate action groups soon followed the direction of 350.org. Jamie Margolin (2020, p. 24), the teenage founder of Zero Hour, for example, wrote her *Youth to Power* book as a "manifesto of the youth revolution" and "guide to causing good trouble, unlearning everything you've been taught before, disrupting the status quo, making your voice heard, challenging problematic authority, changing the culture, changing laws, and yes, changing the world." Drawing on her own experience as well as that of dozens of other youth activists, Margolin offers lessons on "Movement and Organization Building 101," "Event Organizing 101," "Civil Disobedience for Youth 101," "Lobbying as a Young Person 101," "How to Create a Media Strategy 101," and "Social Media Activism 101."

In 2015, a group of graduates from the university fossil fuel divestment movement co-founded the Sunrise Movement in the United States, as a vehicle for "building an army of young people to make climate change an urgent priority across America, end the corrupting influence of fossil fuel executives on our politics, and elect leaders who stand up for the health and wellbeing of all people" (Sunrise Movement, 2021a). Best known for promoting the proposed Green New Deal legislation and working to elect politicians who will support the Green New Deal, the Sunrise Movement, like 350.org, operates with an explicit "theory of change." This is based on building "people power"—developing "an active base of public support by having person-to-person conversations with our family, friends, neighbors, and communities and drawing people into meaningful moments of action through escalated moral protest"—along with "political power"—catalyzing "a critical mass of enthusiastically supportive public officials who will fight for a shared vision of a just future"—as well as "the people's alignment"—which entails pulling together "an extensive network of movements and groups united by the shared vision of a government that fights for dignity and justice for all" (Sunrise Movement, 2021a). Sunrise Movement activists focus on using the power of storytelling, having a powerful, utopian vision (a pedagogy of hope), linking with other, broad and pressing social justice concerns, meeting people where they are (starting with people's local self interests), employing creative and disruptive direct action tactics, and mobilizing the power of art, music, video and social media as effective communication and engagement tools (Prakash, 2020; Rapid Transition Alliance, 2021; Willcox & Barnett-Loro, 2019).

Most recently, the creation of Extinction Rebellion in the United Kingdom in 2018 and its subsequent spread to forty five countries around the world constitutes a third sector of the climate movement that has promoted a close focus on the question of social power (Gunningham, 2019). Marked by its embrace of sustained, mass acts of civil disobedience that seek to disrupt and shut down public spaces in large urban centers for days on end, Extinction Rebellion also offers an explicit theory of social power and change. As articulated by one of its co-founders, Roger Hallam, the theory is called the "civil resistance model" for how to successfully push for radical change. According to Hallam (2019, pp. 179–180), decades of social movement experience and research show that "conventional campaigning methods" based on "persuasion and information" are ineffective in bringing about "radical system change." Instead, what is needed are acts of non-violent disruption. Transformative social power requires the direct participation of large numbers of people—Hallam estimates this to require several tens of thousands of participants, or approximately 1–3 per cent of the national population—to go to the capital city and break the law for an extended period of time—"one or two weeks on average"—by holding a non-violent, fun and engaging mass action that significantly disrupts the daily functioning of that city (pp. 181–188). As Oscar Berglund and Daniel Schmidt (Berglund & Schmidt, 2020, pp. 4, 79) note, the fact that Extinction Rebellion have such a clearly articulated theory of social power and change "has galvanised support for the movement" and helped to "pull in new activists."

In all of these sectors of the climate movement, there have been intense debates and critiques over the theories and strategies of building movement power that have been embraced by different climate action groups. Some question the effectiveness and impact of 350.org's attempt to use the symbolic and economic power of the university and fossil fuel divestment as a climate action model (Bergman, 2018; Gunningham, 2019). Extinction Rebellion has been widely criticized for its core "civil resistance model," its attempt to avoid questions of politics, ideology and political economy, and its failure to address the injustices of racism, colonialism and economic inequality (Alberro, 2020; Berglund & Schmidt, 2020; Carver, 2020). But the key point to take note of is not so much which is the most accurate and effective theory of social power for addressing the climate crisis, but the fact that questions of the relationship between social power, collective action and transformative social change have become central in key sectors of the climate movement—in marked contrast to what has been happening in the field of (formal) climate change education.

Indeed, there is now a cottage industry of advisers, experts and consultants who conduct research, produce reports and run trainings for climate activists on how to grow social power and take effective collective action. In the United States, the Climate Advocacy Lab was set up in

2015 with the goal of "helping the climate community build grassroots power and win through evidence-based advocacy" (Climate Advocacy Lab, 2021). The deputy director of the Climate Advocacy Lab, Carina Barnett-Loro, co-authored an academic article that calls for more systematic research on how the climate movement can build "political power needed to effect meaningful change" (Han & Barnett-Loro, 2018, p. 1). Other organizations, such as the Innovation Network's Social Movement Learning Project, the Mobilisation Lab and Mindworks (both initially set up by Greenpeace), Rapid Transition Alliance, SmartMeme Studios, the New Economy Organisers Network (based in the UK) and many others, offer similar kinds of services. At the Sunrise Movement, many of the core leadership team developed their approach to power and collective action by taking trainings with a group called Momentum (Kingkade, 2019; Leber, 2019). Momentum is "a training institute and movement incubator," established in 2014, that seeks to "give progressive organizers the tools and frameworks to build massive, decentralized social movements" (Momentum, 2021).

Social Movement Learning and Education for Social Change

Much of this research, discussion and practice of social power and collective action in the climate movement is framed as being about activism, organizing, movement building and politics, rather than being about education *per se*. Yet, all of this is also an example of what is sometimes referred to as social movement learning. While social movements have been largely ignored or marginalized in the field of educational studies, there is a strong case for arguing that social movements and education are vital to one another and deeply connected (Niesz et al., 2018). Social movements—which include things like the US civil rights movement, suffragette movement, LGBT rights movement, early labor movement, anti-apartheid movement and so on—have been defined "as temporary public spaces, as moments of collective creation that provide societies with ideas, identities, and even ideals," and as organized, large scale and temporally continuous collective actions that "influence beliefs, practices, and identities; create new networks, organizations, and communities; and transform aspects of the broader culture" (Eyerman & Jamison, 1991, p. 4; Van Dyke & Taylor, 2019, p. 483). Education and learning are thus at the heart of what social movements do, and social movements need to be understood as fundamentally educational or pedagogical institutions (Tarlau, 2014). As Tricia Niesz (2018, p. 2) and her co-authors point out, "educational processes and contexts are crucial to the ways in which social movements' ideas, identities, and ideals are generated and promoted, taught and learned, contested and transformed."

Social movements are also defined as being about social power and social change. While conservative or reactionary social movements seek to prevent change and preserve established systems of power, progressive social movements aim to effect social change and transform power structures. "Fostering or halting change is the *raison d'être* for all social movements," David Snow (2019, p. 7) and his co-authors observe. Focusing on progressive movements, Edwin Amenta and his colleagues (Amenta et al., 2010, p. 288) define social movements as "actors and organizations seeking to alter power deficits and to effect social transformations … by mobilizing regular citizens"; while Phyllis Cunningham (1998, p. 20) similarly describes social movements as "political sites for redistributing power and devising more equitable social structures" (quoted in Rule, 2011, p. 281). What this means is that (progressive) social movements are the one place in which we can find education for social change and learning power being explicitly and consistently embraced—in contrast to formal education, where the idea of education for social change is often considered to be difficult, problematic or inappropriate, and "too political" to be done in school. If we want to think and learn about education for building power and effecting social change, we need to look closely at the long, rich heritage of social movement learning. Indeed, what we find in the recent turn to power in the climate movement is the embrace of deliberate study of the history of social movement education: the ideas and practices about social power and change that we find in Blockadia, and organizations like 350.org, the Sunrise Movement, and Extinction Rebellion are not newly invented, but have been adopted through direct engagement with the lessons of social movements past. If climate change activists are currently seeking to learn from previous traditions of social movement education, there is a strong argument to be made that climate change educators need to do the same.

Beyond the specific claims about building power and developing action that we find now in the climate movement, there at least three key components for developing effective education for social change we can take from the field of social movement learning, along with a host of other invaluable insights. The first is that progressive social movements have consistently developed their own parallel and independent educational institutions, outside and beyond the formal school system, as vital spaces for supporting effective learning for social change. Myles Horton, who founded the independent Highlander Folk School, that played a central role in supporting early industrial labor organizing in the United States as well as the civil rights movement, explains the reasoning behind this strategy:

> We deliberately chose to do our education outside the schooling system. … We concluded that reform within the system *reinforced* the system, or was co-opted by the system. Reformers didn't change the system, they made it more palatable and justified it, made it more humane, more intelligent. … [W]e decided we'd work outside the system and be completely free to do what we thought was the right thing to do in terms of the goals that we set for ourselves and the people we were working for. … We said we could go further in trying to experiment … because we didn't have to conform to anything. Nobody could tell us what to do. We could make our own mistakes, invent our own process. (Horton & Freire, 1990, pp. 199–200)

Examples of social movement educational institutions include not only Highlander, but the Freedom Schools and Citizenship Schools of the US civil rights movement; the American Labor Colleges of the early twentieth century; Sunday Socialist schools in the US and Britain in the early twentieth century; Black Supplementary Schools in Britain; the schools of the Landless Workers Movement in Brazil (Edwards & McCarthy, 1992; Isaac et al., 2020; Niesz et al., 2018). The nonformal educational institutions and spaces created by social movements range from whole schools to one-off training workshops, and may take on any number of different forms: residential schools, summer schools, evening and weekend classes, study circles, reading groups and so on. The value of such spaces is both that they provide the opportunity for movement participants to undertake deliberate, systematic study, learning, reflection and discussion; and that, as Horton suggests above, this can be done without the need to confront the kinds of constraints on educational pedagogy, curriculum, assessment, governance, purpose and access that are often encountered by teachers and learners in formal education institutions.

We can see a proliferation of these kinds of nonformal educational spaces in the climate movement today. In 2016, Indigenous educators and protesters opposing the Dakota Access Pipeline in the US established the Defenders of the Water School, based directly on the model of the Freedom Schools of the US civil rights movement (Eagle Shield et al., 2020). The Sunrise Movement has also created a number of different educational spaces. These include short "Sunrise School" programs that typically run for three weeks, meeting twice a week. In Movement Building 101, participants "learn about the crises gripping our society and how to confront them, build community with others around you who are ready to make lasting social change," and work with "a small group of people from your areas, taking meaningful action together and building power together where you live" (Sunrise Movement, 2021b). Sunrise Creative School, meanwhile, is an "art school for the movement," in which participants learn how to use graphic design, visual art and video in ways that "makes our message clear, shows the power of our actions, and grows our movement" (Sunrise Movement, 2021c). In 2018, Sunrise ran a longer, intensive "Sunrise Semester," as a six month fellowship program for participants "to learn the ropes of electoral campaigning and social movement organizing all while exposing and ending Big Oil's corruption of our political process and making climate change a decisive issue in the 2018 elections" (Negron, 2018). Fellows began by attending a week-long residential school at Camp Galil in rural Pennsylvania—"Sunrise Semester Bootcamp"—before spreading out across the country to work on local political campaigns. Reading the accounts of bootcamp sessions suggests a strong similarity with residential programs run at Highlander Folk School by Myles Horton, Septima Clark and others decades earlier (Adams & Horton, 1975; Negron, 2018; Sunrise Movement, 2018a, 2018b).

A second core component of social movement learning is that, though creating intentional, nonformal educational spaces is important, informal learning is at the heart of this kind of learning. According to Griff Foley (1999, p. 3), learning in social movements "is largely informal and often incidental—it is tacit, embedded in action and is often not recognised as learning." Through participating in social movement organizing and action, individuals learn "about issues central to the social movement," "the vision of the movement," and "skills and practices of organization, mobilization, and collective action"; they also learn about "individual and collective identity," or "who they are and who they are becoming," and develop their own "social critique and agency" (Niesz et al., 2018, p. 15; see also Choudry, 2015; Rule, 2011). Jonathan Langdon (2011, p. 155) differentiates between "learning *in* struggle"—referring to "the long-term incidental learning that takes place within a struggle over time"—as opposed to "learning *through* struggle"—which is "the concentrated learning that is produced during the course of a particular engagement, action, or even campaign"—and "learning *to* struggle"—that concerns the normative beliefs that emerge in a movement of the most effective ways to build power and achieve social change.

A central argument in this literature is that there is something about learning about power, action and social change through direct experience that cannot easily be replicated in formal classroom contexts; and therefore, one central importance of social movement organizing and action is to enable this kind of learning to occur. As Foley (2001, p 0.78) writes of participants in women's community center organizing in Australia:

> Although [organizing activities] were often painful for the individuals involved … they also generated what is probably the most significant sort of human learning. This is learning that enables people to make sense of and act on their environment, and to come to understand themselves as knowledge-creating, acting beings. … [W]omen learned to overcome … fear and lack of confidence, … to fight for something for themselves and to participate in difficult collective decision-making.

Foley is effectively describing a process that is more widely known through Paulo Freire's (2000, p. 160) concept of conscientization, in which "people, through a true praxis, leave behind the status of *objects* to assume the status of historical *Subjects*" (emphasis in original). A recent example of this kind of learning can be found in Eric Blanc's (2019) *Red State Revolt*, his account of the wave of successful teacher labor organizing and strike action that swept across the United States in 2018. Blanc talks of the teacher strikes as themselves being a form of pedagogy, arguing that "most working-class people learn about social power through their experiences in struggle and mass organizing" (p. 157). "A book or a lecture cannot effectively explain, on a mass scale, ideas like solidarity or collective action," writes Blanc, "for these to sink in, you need to experience them firsthand" (pp. 157–158). "More than anything, the strike changed people's ideas of what is possible," says one striking teacher in West Virginia (pp. 158–159). "We learned that … you just have to have courage and a backbone to stand up for yourself, for your state, and your kids," says another: "And that's what we did" (p. 145).

A third key component of social movement learning is that it is concerned not just with internal learning of participants, but external learning of the general public and political and economic elites as well. As Aziz Choudry (2015, p. 8) notes, "social movements can … play a major educative role for broader publics who are not engaged in them." In fact, a primary focus of social movements is to create actions and events that can stimulate and shape what is variously called "societal learning" or "public learning," that may encompass the aims of "informing the public" or "raising public awareness of matters related to the public interest," producing "changes in worldview" and "society's core values," and spurring the public to take action on the movement's core issues (Krajnc, 2000, pp. 341–342). Social movements seek to shape public learning through a variety of methods, including cultural productions (music, visual art, film, theatre, literary fiction, poetry, etc.), independent media (online and/or print news outlets, magazines, books and leaflets), or staging public events and collective actions (Krajnc, 2000; Niesz et al., 2018). A strike, a picket line, a march, a protest, a demonstration, an occupation, after all, is intended to be a form of pedagogy, not just for movement participants, but for local and more dispersed public audiences as well. As

Budd Hall et al. (2006, p. 7) write, in their review of social movement learning:

> The actions of social movements, be they large scale media events such as Greenpeace and other environmental groups have staged, or benefit concerts for victims of HIV/AIDS, or the creation of quilts by women to protest the building of an unwanted power station on Vancouver Island …, create rich environments for learning by large numbers of the public.

As social movements grow larger and more powerful, their "repertoire of cultural artefacts can be assimilated into the wider cultural field, to contribute towards a new structure of feeling" in society more generally—and they can come to have the large scale educational, cultural and transformative impacts on public beliefs, identities and practices that the most influential social movements in history are recognized for having accomplished (McGregor et al., 2019, p. 502).

Beyond these general components, we can find all kinds of essential insights into ways of doing effective education for social change and learning power by examining the educational practices of specific social movements. For example, if we consider the schools involved with the US civil rights movement—Highlander Folk School, the Citizenship Schools and Freedom Schools—we can find a strong commitment to holistic approaches to social change education, that seek to move not just the head, but the heart and hands as well. "We tried to involve everybody in singing and doing drama and dancing and laughing and telling stories because that's a part of their life," Myles Horton (2003, p. xii) says of the social change workshops run at Highlander, because "the way people live was more important than any class or subject we were dealing with," and provided a powerful "learning experience" that stimulated "senses other than their minds." We find recognition of the importance of educational embedding. Curriculum designers for the Freedom Schools paid attention to the smallest details, even the pronouns used in class discussions: asking students to think not about what "I want" individually, but what "we want" collectively as African-American youth, on the belief that this mattered for shaping the kind of learning that would best support the civil rights movement (Chilcoat & Ligon, 1998, p. 175). But equally, civil rights movement educators also recognized that details of classroom practice took on powerful effects for students because they were embedded within and linked to wider movement goals and activities. Teachers and students in the Freedom Schools had a clear sense of collective purpose, and participated in civil rights movement actions, from voter registration to protesting local institutional segregation. It is this embedding of classroom education within broader social movement structures, goals and actions that Rebecca Tarlau (2014, p. 371) argues has been lost in recent school based social justice education, as "critical pedagogy has largely been disconnected from its organizing roots." A third commitment we find in the civil rights movement schools is the shunning of what Lilia Bartolomé (1994) calls "the methods fetish," or belief there is one correct form of pedagogy or curriculum that will guarantee successful realization of movement goals. Instead, the conviction was that movement educators needed to constantly reflect on and adapt their practices to changing goals and contexts (Chilcoat & Ligon, 1998). "Some observers have tried to isolate a Highlander methodology or technique," Myles Horton (2003, pp. xiii, 178) reflects, but "I'm less interested in methodology or techniques than I am in a *process* that involves the total person, involves vision, involves total realities." "Techniques are something you use and discard," Horton argues, "Let's minimize our loyalty to institutions that do not serve the needs of people" (p. 178).

The Cycle of Community Organizing and the Popular Education Spiral

In seeking to conceptualize the strategies and approaches for learning power and collective action that can be found in social movement and political organizing campaigns, there are two models of radical education that are often drawn upon, whether explicitly or implicitly. One is the cycle of community organizing (see Fig. 8.1), that was discussed by the students from Notre Dame Girls' School at the beginning of the chapter, and is linked with Alinskyite institutions and traditions (Macleod & Whelan, 2015; Tattersall, 2015). The second is the popular education

Fig. 8.1 The cycle of community organizing. Based on training materials from Citizens UK, the Sydney Alliance and PICO: See Bolton, 2017; Shannahan 2014; Tattersall, 2015

spiral (see Fig. 8.2), based on the ideas of Paulo Freire, that was developed by social justice educators working at the Doris Marshall Institute for Education and Action in Toronto, and adopted and popularized by the Highlander Folk School (Arnold et al., 1991; Gilbert, 2021; Spence, 2015). While many campaigning groups use some variation or combination of these models, it can be helpful to consider the similarities and differences of these two models, particularly as these relate to power and action. For each model focuses on a slightly different set of priority

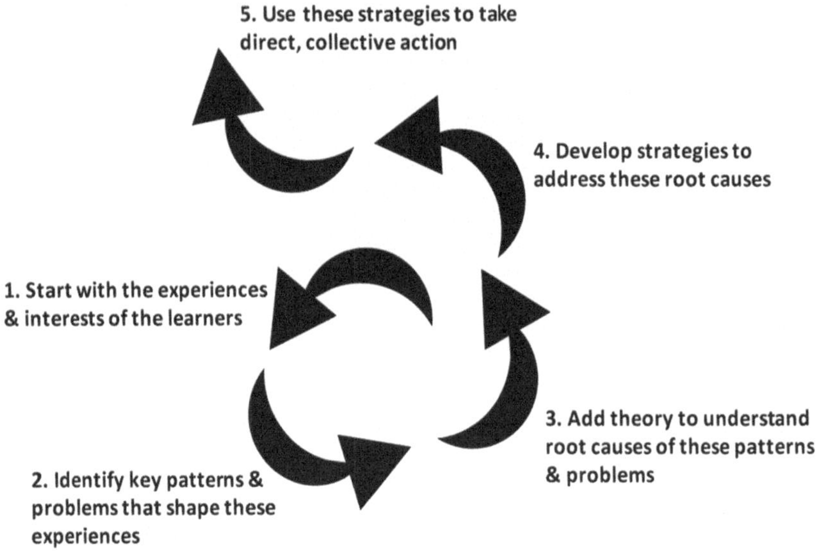

Fig. 8.2 The popular education spiral. Based on training materials found in Arnold et al., 1991; Gilbert, 2021; Spence, 2015

concerns; and together, these models can offer a clear, theoretical and practical guide for how to foreground concerns with learning power and collective action in climate change education.

Both the cycle of community organizing and the popular education spiral offer three components that differentiate them from many other models of social justice education. Both insist that the vital starting point is listening to and learning from local community members, in order to ground and shape the educational agenda and process. In community organizing, this is done through one-to-one relational meetings and listening campaigns (Chambers, 2018); while in popular education, Freire speaks of the importance of conducting preliminary anthropological research. Popular education "starts with the conviction that it cannot present its own program but must search for this program dialogically with the people," writes Freire (2000, p. 124). It does this through a process of "research [that] is carried out during informal encounters with the inhabitants of the area," that can "reveal longings, frustrations, disbeliefs, hopes, and an impetus to participate" (Freire, 1974, p. 46). This initial step is essential for addressing the gap that often exists between local concerns and interests, and educational curriculum and pedagogy. "One may have the finest teachers, the best libraries, the most beautiful buildings," Alinsky (1969, p. 178) writes, "but unless the people have a desire to use these facilities all of the teachers, buildings and libraries will not advance the cause of education." From this first step, both models focus on identifying key social problems in the lives of community members, through a process of discussion, further research and critical reflection. Both community organizing and popular education go beyond the commitment to child- or student-centered learning found in progressive education, in that this step is interested in learning about collective more than individual interests and concerns, and is understood as requiring an active and extended process of listening or research.

From these initial steps of listening, learning and reflection on core social problems, both models progress to an essential final step that is also radically different from many other social justice education frameworks: a move from learning toward taking strategic, collective action to directly address social problems. As Freire (2000, p. 87) writes:

> When a word is deprived of its dimension of action, reflection automatically suffers as well; and the word is changed into idle chatter, into *verbalism*, into an alienated and alienating 'blah.' It becomes an empty word, one which cannot denounce the world, for denunciation is impossible without a commitment to transform, and there is no transformation without action. (Emphasis in original)

Community organizing likewise focuses on the central importance of moving to take action (Chambers, 2018). A community "organization needs action as an individual needs oxygen," writes Alinsky (1971, p. 87), for without sustained and effective action, the organization will eventually wither and die. Indeed, what is referred to here as the cycle of community organizing is also known as the "cycle of action," as it is action that is seen as being central to this entire learning process. In focusing on the question of action, the popular education and community organizing approaches insist that the link between education and social transformation is made an explicit concern for educators and learners. For very often, this is the link that is left vague and unspecified in social justice education: how exactly is it that education is supposed to lead to radical social transformation? In other traditions, the creation of prefigurative utopian spaces in the classroom and school, and changes in ideas and attitudes among learners are expected and intended to lead to broader changes in society somewhere further along the road; but the exact process for how this is going to happen is not entirely clear.

Finally, in focusing on this essential link between education and action, both popular education and community organizing are centrally concerned with questions of how to build social power in local communities, and transform power imbalances in society. This is at the heart of how community organizing is defined. "Community organizing seeks to *alter the relations of power* between the groups who have traditionally controlled our society and the residents of marginalized communities," write Schutz and Sandy (2011, p. 12; emphasis in original). For Freire and other popular educators, the primary aim of popular or liberatory education is to overcome local, regional and global structures of oppression. "The issue of power is central," writes Rick Arnold (1991, p. 1) and the social justice educators at the Doris Marshall Institute who developed the popular education spiral: "Education must empower all people to act for change," so they can "unmask" and "transform" the "power relations in our society." Educators working in both of these traditions have subsequently developed a range of different "power analysis" tools and frameworks to help learners understand, research and contest structures of social power in their communities (see, for example, Bolton, 2017; McGee & Pettit, 2020; Schutz, 2019).

However, while being alike in these respects, the community organizing cycle and popular education spiral are quite different in other regards: and one vital difference is that they tend to focus on different aspects of power in society. Social power is commonly said to have different "faces," "types" or "dimensions." One of the most recognized models for talking about different dimensions of social power, based on the work of Lukes (2005) and JASS (Miller et al., 2006), distinguishes between visible, invisible and hidden power: *visible power* refers to the "public aspects of political power—the formal rules, authorities, institutions and procedures of formal decision-making and enforcing the rules"; *hidden power* exists where "powerful people" and "organised interests" seek to use "their influence behind the scenes to control who gets to the decision-making table and what concerns shape the public agenda"; while *invisible power* is about "the power of beliefs, ideology, social norms and culture to shape people's worldview, sense of self, values and acceptance of what is normal, right and even real" (Bradley, 2020, pp. 104–105). Social theorists and activists often point to invisible power as "probably the most insidious of the three dimensions of power," for the reason that it:

> shapes the psychological and ideological boundaries of change. Significant problems and ideas are not only kept from the decision-making table, but also from the minds and consciousness of the people involved, even those directly affected by the problem. … Processes of socialization, culture and ideology perpetuate exclusion and inequality by defining what is normal, 'true,' and acceptable. These processes also operate in ways to make injustices like poverty, racism, sexism and corruption invisible to the society at large, and make those who experience systematic discrimination the object of blame, including blaming themselves. (Miller et al., 2006, p. 12)

If we are to address the climate crisis effectively, we need to learn how to work with all of these forms of visible, hidden and invisible power: to learn how to challenge official decision makers, transform political agendas, and shift cultural discourses and ideologies.

The cycle of community organizing tends to focus on the first two dimensions of power, as it often conceptualizes power as "the capacity to influence (or affect) the actions of powerful people and institutions" (Schutz & Sandy, 2011, p. 23). Much of the research, action and evaluation in the community organizing cycle focuses on learning how to impact political agendas and official decision makers. Learners learn the nuts and bolts of how to identify key social problems, carve these into smaller issues, and develop actionable demands that can address these issues. They learn how to build strong relationships in local communities that can help to marshal collective grassroots power. They learn how to identify targets who can address their demands, and other key actors who may support or oppose their campaigns. They learn how to develop effective actions that can help them to win campaigns and build collective power (see, for example, Bolton, 2017; Chambers, 2018; Schutz & Sandy, 2011). Learning power in the cycle of community organizing is practical, pragmatic and skills based. Discussions of ideology are notoriously avoided in Alinskyite approaches, and systematic political education and consciousness raising projects are often missing (Sen, 2003). Community organizations that work in the Alinsky tradition tend to believe "they always have to have an issue that is on the front burner," argue DeFilippis, Fisher and Shragge (2010), and one "consequence is that the educational dimensions … that contribute to a long-term vision, can become lost."

The popular education spiral, on the other hand, focuses partly on the question of hidden power, but is principally concerned with contesting invisible power (Schutz, 2019, p. 96). Compared to the cycle of community action, the spiral adds in an extra step between its initial community-based research and its move to collective action, to develop "theory linked to the patterns in what people know." This is about using education to foster what is known in the Freireian tradition as "conscientization" or "awakening of critical awareness" (Freire, 1974, p. 15); and conscientization is understood to be a vital step in the path toward radical social transformation. Learners need to understand the deep social structures and cultural discourses that shape the local conditions of their own existence—for example, the social histories of racism, patriarchy, capitalism and colonialism. To do this, they engage in what Rinku Sen (1983) and others call "political education," that seeks to critically examine "the core ideas that shape society":

> [W]e need to be clear and vocal about what we believe, about the basis on which we oppose economic and social policies, and about the kinds of systemic changes we want. … To be more ideologically ambitious, we have to engage in analysis and political education. We have to read, share information, understand history, bring people to speak to our groups. … We have to think about our theories of how society is organized, why it is organized that way, and how change will come. We have to be willing to integrate our experiences with information because no single person can experience everything.

This process of liberatory or problem-posing or consciousness raising education is internal, focused on transforming the social imagination and self-understanding of local community members, so that they can come to see exactly why and how their society could and should be changed. "A deepened consciousness of their situation leads people to apprehend that situation as an historical reality susceptible of transformation," Freire (2000, p. x) argues, and "resignation gives way to the drive for transformation and inquiry." But in the context of social movement learning and action, this process is also external, as it seeks to challenge and shift broader cultural logics and common sense frameworks for understanding the world that prevail among the general public as well. This process of conscientization might encompass anything and everything: from unlearning the sense of learned helplessness and fear of taking action that often prevails among marginalized communities; to developing a sophisticated analysis of global capitalist economy; to using this analysis to raise critical public awareness of the links between consumption patterns, the political agendas and self interests of multinational corporations, and the ongoing destruction of local environments and livelihoods worldwide (for an excellent example of this kind of popular education, see the discussions of the organizing campaigns of the Coalition of Immokalee Workers in Bell, 2007; Haedicke, 2020; Marquis, 2017; and Soltis, 2006).

Historically, the approaches taken by the cycle of community organizing and popular education spiral have often been seen as in conflict with one another. But recently, there has been a growing sense among many educators, researchers and activists that the insights of both models are needed to develop effective social change campaigns (DeFilippis et al., 2010; Engler & Engler, 2016; Walls, 2015). Dave Beck and Rod Purcell (2013) argue that "community organizing as a practice is strong in terms of its tactics, identification of local issues and mobilising people for change," but "less strong in the development of critical thinking" and situating "its practice within broader political thinking and political movements." This "lack of criticality," the authors suggest, can be "addressed by forms of practice which take community organising and weave in the thinking of Freire," in order to "develop critical consciousness essential for sustainable social action," and give "a theoretical basis on which to build trans-local networks that have the potential to challenge the current dominant order." Rinku Sen (2003), conversely, warns that "some feminists and racial justice organizers … have reacted to Alinskyist limitations" by "moving too far" in the other direction, and "creating programs that are heavy on … political education but light on campaigns and action." Sen argues instead that "the beauty of innovation in organizing emerges from the marriage of the two: political education creates the reflection and growth opportunities that motivate action, and action provides the expression of newly clarified values." A different way to articulate this point is to argue that effective climate change (and other social change) education not only needs to ensure that learning power and collective action is at the center of its educational agenda: it needs to avoid "superficial approaches to power," by recognizing that "power is dynamic, relational and multidimensional" (Miller et al., 2006, p. 6). As Valerie Miller (2006, p. 6) and her co-authors argue, "failure to deal with the complexities of power can lead to missed opportunities and poor strategic choices," and "can be risky and counterproductive."

Learning Power in Schools: Lessons from Movements for Educational Justice?

Learning power and collective action are thus to be found mostly in the context of social movement activism and in the learning models of community organizing and popular education—neither of which have a strong presence in the formal education sector (Schutz, 2019). Does this mean that to develop climate change education that addresses questions of social power and collective action, we need to turn our backs on schools, universities and other institutions of formal education? This would be a problematic conclusion to make. As the story of Notre Dame Girls' School at the outset of this chapter suggests, this clearly is not a necessary response: in certain contexts, at least, it is possible for schools to embrace these kinds of educational approaches. It is also not a strategic nor sufficient response. As Rebecca Tarlau argued in the Introduction to this book, the formal education sector remains an essential site for social change education of all kinds, even if there is a significant amount of social movement activism and non-formal education occurring as well. The reason, as Tarlau (2019) suggests, is the universal, compulsory and mass nature of formal education, as well as the significant financial and material resources invested in the sector by the state and other actors. Schools, colleges and universities engage with and educate children, youth and young adults on a regular, often daily, and continuing, long term basis, in ways that few social movements can hope to replicate. Schools and other institutions of formal education matter, too, since this is where officially sanctioned knowledge, values, ideologies and beliefs are reproduced.

In fact, as the work of Tarlau and others has demonstrated, there is another vital link between social movements and education that is important to recognize. Social movements have long sought to shape and reshape schools, universities and the formal education sector, as an essential terrain of cultural, political and ideological struggle (Niesz et al., 2018). As Jean Anyon (2009, p. 198) observes, in the United States, many progressive changes in formal education over the past century come directly from social movement organizing:

> Immigrant and labor organizations' struggles for education in the late 19th and early 20th centuries yielded adult worker education and Catholic schools for children; the Civil Rights Movement led to national Head Start programs as well as increased recognition and educational opportunities nationwide; Latino struggles produced Bilingual education; the 1970s women's movement yielded curricular change as well as increased entitlements in schools and districts; disabilities organizing also has prompted federal protections and entitlements.

Increasingly, many researchers, theorists and activists have come to recognize that radical changes in the formal education sector are most likely to happen, not through top down policy interventions, but through the power of social movement organizing and collective struggle (Oakes & Rogers, 2006; Warren, 2014). "It is social movements that provide the engines of lasting educational transformation," argues Michael Apple (2003; quoted in Niesz et al., 2018, p. 18). In the United States and other countries, there has been a growth of local community organizing for education reform, and a broad-based national educational justice movement that seeks to learn from past struggles in order to radically transform schools for the future (Ferman, 2017; Su, 2009; Warren, 2018).

This is a movement that holds great promise for bringing a focus on learning power to schools and universities, and thereby, support the development of effective climate change and social justice education. Social movements for education justice and reform provide key spaces for students, teachers, parents and other community members to learn about social power and collective action directly, through participating in these movements. But movements for education justice and reform also mobilize the kind of grassroots community power that can push for radical changes in the central purpose and practice of formal education: to open schools up to concerns with learning power and social change, address fundamental social problems like the climate crisis, and rethink limitations of the "core grammar" of schooling (discussed in Chap. 2) (Oakes & Rogers, 2006; Warren, 2014). As suggested at the beginning of this book, in developing effective climate change education, we should never be led into thinking we have to choose between either social movement learning and independent, non-formal education projects, or schools and other formal educational institutions. On the contrary, the aim needs to be to think through how all of these different kinds of learning spaces and institutions can be marshalled together, in ways that can create powerful and progressive social change. One of the insights of the Sunrise Movement and other supporters of the Green New Deal in the United States is that to be effective, the climate change movement needs to link more closely with other social justice movements—with the struggle against racism, colonialism and patriarchy, for example, and collective demands for good quality homes, jobs and health care. A parallel argument can be made about climate change education. To develop effective climate change education, it is likely that climate change educators need to link with broader, social movement struggles for education justice overall, and with collective projects that seek to develop sweeping, utopian visions—pedagogies of hope—for radically transforming the education system as a whole.

References

350.org. (2019). Climate change is about power. https://clairepinegar.com/350-org-Theory-of-Change

350.org. (2021). 350 theory of change. https://350ct.org/350-theory-of-change/

Adams, F., & Horton, M. (1975). *Unearthing seeds of fire*. Blair.

Alberro, H. (2020). Extinction Rebellion: Why disavowing politics is a dead end for climate action. *The Conversation*, 9 September. https://theconversation.com/extinction-rebellion-why-disavowing-politics-is-a-dead-end-for-climate-action-145479

Alinsky, S. (1969). *Reveille for radicals*. Random House.

Alinsky, S. (1971). *Rules for radicals: A pragmatic primer for realistic radicals*. Vintage Books.

Amenta, E., et al. (2010). The political consequences of social movements. *Annual Review of Sociology, 36*, 287–307.

Anyon, J. (2009). Progressive social movements and educational equity. *Educational Policy, 23*(1), 194–215.

Apple, M. (2003). What effects do social movements have on curricular reform? *Educational Policy, 17*(4), 519–525.

Arkwright, C. (2020). Positive images: How young people organised to change policy and build better relationships with the police. In S. Chapleau (Ed.), *Schools in their communities* (pp. 36–39). Big Education.

Arnold, R., et al. (1991). *Educating for a change*. Between the Lines.

Bartolomé, L. (1994). Beyond the methods fetish. *Harvard Educational Review, 64*(2), 173–195.

Beck, D., & Purcell, R. (2013). *International community organising: Taking power, making change*. Policy Press.

Bell, B. (2007). Florida farmworkers build unity through education and action. *Race, Poverty & the Environment, 14*(2), 39–41.

Berglund, O., & Schmidt, D. (2020). *Extinction Rebellion and climate change activism*. Palgrave Macmillan.

Bergman, N. (2018). Impacts of the fossil fuel divestment movement. *Sustainability, 10*, 1–18.

Blanc, E. (2019). *Red state revolt*. Verso.

Bolton, M. (2017). *How to resist: Turn protest to power*. Bloomsbury.

Boström, M., et al. (2018). Conditions for transformative learning for sustainable development: A theoretical review and approach. *Sustainability, 10*(4479), 1–21.
Bradley, A. (2020). Did we forget about power? Reintroducing concepts of power for justice, equality and peace. In R. McGee & J. Pettit (Eds.), *Power, empowerment and social change*. Routledge.
Bradshaw, E. (2015). Blockadia rising. *Critical Criminology, 23*, 433–448.
Carver, E. (2020). To force climate action, we need more than just protests. *Jacobin*, 11 December. https://www.jacobinmag.com/2020/12/climate-change-protest-strategy-electoral-politics-sunrise-extinction-rebellion
Chambers, E. (2018). *Roots for radicals: Organizing for power, action and justice*. Bloomsbury.
Chilcoat, G., & Ligon, J. (1998). 'We talk here. This is a school for talking': Participatory democracy from the classroom out into the community: How discussion was used in the Mississippi Freedom Schools. *Curriculum Inquiry, 28*(2), 165–193.
Choudry, A. (2015). *Learning activism*. University of Toronto Press.
Climate Advocacy Lab. (2021). About. https://climateadvocacylab.org/about-public
Cunningham, P. (1998). The social dimension of transformative learning. *PAACE Journal of Lifelong Learning, 7*, 15–28.
D'Arcy, S., et al. (2014). *A line in the tar sands*. Between the Lines.
DeFilippis, J., Fisher, R., & Shragge, E. (2010). *Contesting community*. Rutgers University Press.
Eagle Shield, A., et al. (Eds.). (2020). *Education in movement spaces*. Routledge.
Edwards, B., & McCarthy, J. (1992). Social movement schools. *Sociological Forum, 7*(3), 541–550.
Engler, M., & Engler, P. (2016). *This is an uprising*. Bold Type Books.
Eyerman, R., & Jamison, A. (1991). *Social movements: A cognitive approach*. Pennsylvania State University Press.
Ferman, B. (Ed.). (2017). *The fight for America's schools*. Harvard Education Press.
Foley, G. (1999). *Learning in social action*. Zed.
Foley, G. (2001). Radical adult education and learning. *International Journal of Lifelong Education, 20*(1), 71–88.
Freire, P. (1974). *Education for critical consciousness*. Bloomsbury.
Freire, P. (2000). *Pedagogy of the oppressed*. Continuum.
Gilbert, C. (2021). Walking the popular education spiral. *Educational Action Research*. https://doi.org/10.1080/09650792.2021.1875856

Gunningham, N. (2019). Averting climate catastrophe: Environmental activism, Extinction Rebellion and coalitions of influence. *King's Law Journal, 30*(2), 194–202.

Haedicke, S. (2020). Coalition of Immokalee Workers: Farmworker-led popular education and performance. *Research in Drama Education, 25*(4), 576–580.

Hall, B., et al. (2006). *Social movement learning.* University of British Columbia.

Hallam, R. (2019). The civil resistance model. In C. Farrell et al. (Eds.), *This is not a drill* (pp. 178–191). Penguin.

Han, H., & Barnett-Loro, C. (2018). To support a stronger climate movement, focus research on building collective power. *Frontiers in Communication, 3,* 1–5.

Hayes, K. (2020). For water, love, and liberation: We learned together. In A. Eagle Shield et al. (Eds.), *Education in movement spaces.* Routledge.

Henderson, J., & Drewes, A. (2020). Teaching climate change in the United States. In J. Henderson & A. Drewes (Eds.), *Teaching climate change in the United States.* Routledge.

Hestres, L. (2015). Climate change advocacy online. *Environmental Politics, 24*(2), 193–211.

Hestres, L., & Hopke, J. (2020). Fossil fuel divestment: Theories of change, goals, and strategies of a growing climate movement. *Environmental Politics, 29*(3), 371–389.

Horton, M. (2003). *The Myles Horton reader.* Ed. D. Jacobs. University of Tennessee Press.

Horton, M., & Freire, P. (1990). *We make the road by walking: Conversations on education and social change.* Temple University Press.

Isaac, L., et al. (2020). Social movement schools. *Social Movement Studies, 19*(2), 160–182.

Jorgenson, S., Stephens, J., & White, B. (2019). Environmental education in transition: A critical review of recent research on climate change and energy education. *Journal of Environmental Education, 50*(3), 160–171.

Kingkade, T. (2019). These activists are training every movement that matters. *Vice,* 18 November. https://www.vice.com/en/article/8xw3ba/these-activists-are-training-every-movement-that-matters-v26n4

Klein, N. (2014). *This changes everything.* Penguin.

Kluttz, J., & Walter, P. (2018). Conceptualizing learning in the climate justice movement. *Adult Education Quarterly, 68*(2), 91–107.

Krajnc, A. (2000). Popular environmental education: Lessons from the labor and civil rights movements. *New Political Science, 223*, 341–360.

Langdon, J. (2011). Democracy re-examined: Ghanaian social movement learning and the re-articulation of learning in struggle. *Studies in the Education of Adults, 43*(2), 147–163.
Leber, R. (2019). Climate change is finally having a political moment. *Mother Jones*, 2 December. https://www.motherjones.com/politics/2019/12/climate-change-is-finally-having-a-political-moment-thats-no-accident/
LeQuesne, T. (2019). From carbon democracy to carbon rebellion. *Journal of World-Systems Research, 25*(1), 15–27.
Lukes, S. (2005). *Power: A radical view* (2nd ed.). Palgrave Macmillan.
Macleod, J., & Whelan, J. (2015). *People power manual: Community organising guide*. Change Agency.
Margolin, J. (2020). *Youth to power*. Hachette.
Marquis, S. (2017). *I am not a tractor! How Florida farmworkers tool on the fast food giants and won*. ILR Press.
Martinez-Alier, J. (2021). *Mapping ecological distribution conflicts: The EJAtlas*. The Extractive Industries and Society https://doi.org/10.1016/j.exis.2021.02.003
McGee, R., & Pettit, J. (Eds.). (2020). *Power, empowerment and social change*. Routledge.
McGregor, C., et al. (2019). Climate justice education: From social movement learning to schooling. In T. Jafry (Ed.), *Routledge handbook of climate justice* (pp. 494–508). Routledge.
McKibban, A., & Steltenpohl, C. (2019). Community organizing, partnerships, and coalition. In L. Jason et al. (Eds.), *Introduction to community psychology* (pp. 286–306). Pressbooks.
Miller, V., et al. (2006). *Making change happen: Power*. Just Associates.
Momentum. (2021). About Momentum. https://www.momentumcommunity.org/about-momentum
Negron, J. (2018). Sunrise semester bootcamp: Day 1! *Sunrise News*. 5 June. https://www.sunrisemovement.org/movement-updates/sunrise-semester-bootcamp-day-1-55aa0952f57/
Niesz, T., et al. (2018). Social movements and educational research. *Teachers College Record, 120*(3), 1–41.
Notre Dame Girls' School. (2019). What is community organising? One school tells us. *YouTube*, 28 February. https://www.youtube.com/watch?v=m6aZ_8P8WZg
Oakes, J., & Rogers, J. (2006). *Learning power: Organising for education and justice*. Teachers College Press.

Prakash, V. (2020). Redefining what's possible. *Sierra*, 22 December. https://www.sierraclub.org/sierra/2021-1-january-february/feature/varshini-prakash-redefining-whats-possible

Rapid Transition Alliance. (2021). The Sunrise Movement. 28 January. https://www.rapidtransition.org/stories/the-sunrise-movement-how-a-us-grassroots-youth-movement-helped-set-the-national-climate-agenda-for-rapid-change/

Rule, P. (2011). Cross-learning: The possibilities of a learning dialogue between the HIV & AIDS and disability movements. *Studies in the Education of Adults, 43*(2), 216–233.

Shannahan, C. (2014). *A theology of community organizing*. Routledge.

Schutz, A. (2019). *Empowerment: A primer*. Routledge.

Schutz, A., & Sandy, M. (2011). *Collective action for social change: An introduction to community organizing*. Palgrave Macmillan.

Sen, R. (2003). *Stir it up: Lessons in community organizing and advocacy*. Jossey-Bass.

Snow, D., et al. (2019). Introduction: Mapping and opening up the terrain. In D. Snow et al. (Eds.), *The Wiley Blackwell companion to social movements* (2nd ed., pp. 1–16). John Wiley & Sons.

Soltis, L. (2006). *Presente! Music, mobilization and global engagement: The case of the Coalition of Immokalee Workers*. PhD dissertation. Emory University.

Speer, P., et al. (1995). Organizing for power: A comparative case study. *Journal of Community Psychology, 23*, 57–73.

Spence, V. (2015). *Where the roads meet: Intersecting perspectives on community literacy*. Master's thesis, University of Tennessee.

Su, C. (2009). *Streetwise for book smarts: Grassroots organizing and education reform in the Bronx*. Cornell University Press.

Sunrise Movement. (2018a). Sunrise semester bootcamp: Days 2–4. *Medium*, 8 June. https://medium.com/sunrisemvmt/sunrise-semester-bootcamp-days-2-4-9d75631a7e39

Sunrise Movement. (2018b). Sunrise semester bootcamp: Days 5–7. *Medium*, 13 June. https://medium.com/sunrisemvmt/sunrise-semester-bootcamp-days-5-7-eae4e1987b8a

Sunrise Movement. (2021a). About the Sunrise Movement. https://www.sunrisemovement.org/about/?ms=AboutTheSunriseMovement

Sunrise Movement. (2021b). Movement building 101. https://www.sunrisemovement.org/movement-building-101/

Sunrise Movement. (2021c). Sunrise creative school. https://www.sunrisemovement.org/sunrise-creative-school/
Tarlau, R. (2014). From a language to a theory of resistance: Critical pedagogy, the limits of 'framing' and social change. *Educational Theory, 64*(4), 369–392.
Tarlau, R. (2019). *Occupying schools, occupying land: How the Landless Workers Movement transformed Brazilian education*. Oxford University Press.
Tattersall, A. (2015). The global spread of community organizing. *Community Development Journal, 50*(3), 380–396.
Temper, L., et al. (2020). Movements shaping climate futures: A systematic mapping of protests against fossil fuel and low-carbon energy projects. *Environmental Research Letters, 15*, 1–23.
Van Dyke, N., & Taylor, V. (2019). The cultural outcomes of social movements. In D. Snow et al. (Eds.), *The Wiley Blackwell companion to social movements* (2nd ed., pp. 482–498). John Wiley & Sons.
Walls, D. (2015). *Community organizing*. Polity Press.
Warren, M. (2014). Transforming public education: The need for an educational justice movement. *New England Journal of Public Policy, 26*(1), 1–16.
Warren, M. (Ed.). (2018). *Lift us up, don't push us out!* Beacon Press.
Willcox, N., & Barnett-Loro, C. (2019). The keys to the Sunrise Movement's momentum? *MobLab*, 13 February. https://mobilisationlab.org/stories/sunrise-movement-bold-idea-visionary-narrative-organising-best-practices/

9
Conclusion

Education has a vital role to play in addressing the climate crisis, as it does in addressing many other social problems and crises. To play this role, it needs at the very least to orient toward the challenge of solving the climate crisis as a key learning aim. As simple as this sounds, this often does not happen, as other aims or goals take priority in schools, universities and other educational institutions around the world, and the problem of the climate crisis is shunted to the side or ignored. Even if the climate crisis is addressed as a legitimate curriculum topic, it can all too easily be framed in limiting and disempowering ways by the core grammar of formal schooling, or made available only to a small minority of already interested students. More than this, however, if education is to play an effective role in helping to address the climate crisis, in needs to do more than just enable learners to understand the nature and causes of this crisis; it needs also to empower learners to be able to make the kinds of fundamental changes in society, culture and economy that are required in order to address the root causes of the climate crisis, and mitigate and adapt to the growing impacts of this crisis. An education that can help to address the climate crisis, in other words, must be an education that is oriented to the project of radical social transformation.

S. Tannock, *Educating for Radical Social Transformation in the Climate Crisis*, Palgrave Studies in Education and the Environment,
https://doi.org/10.1007/978-3-030-83000-7_9

As noted in the Introduction to this book, a small but growing number of scholars and activists are now developing models of what such a form of education should look like, that can help to create "a fundamental transformation of lifestyles and economic patterns [that] is needed to achieve sustainable development" (Balsiger et al., 2017, p. 357). David Selby and Fumiyo Kagawa (2018, p. 302) call for the embrace of "subversive learning" to confront the "key climate change drivers" of "economic growth, consumerism, denial, and climate injustice," as well as "restorative learning" that focuses on "restoring nature intimacy, confronting despair, and reclaiming the good life." A range of scholars call for the adoption of transformative learning, which is defined in various ways, but generally concerns developing "the capacity of individuals and society to transform individual and collective consciousness, social institutions, and economic systems toward planetary boundaries in balance with the carrying capacity of the earth" (Kwauk & Casey, 2021, p. 5; see also Aboytes & Barth, 2020; Boström et al., 2018; Macintyre et al., 2018). An international group of scholars are working collaboratively on developing the concept of "transgressive learning," defined as "a form of transformative learning that addresses structural forms of power" and "generates critical thinking and collective agency and praxis that directly and explicitly challenges those aspects of society that have become normalized, but which require challenging for substantive sustainability transformation to emerge" (Macintyre, 2019, p. 136; Lotz-Sisitka et al., 2016, p. 51; see also T-learning, 2021).

There are many close links and parallels between this emergent body of work and the older set of ideas, practices and traditions of radical education that have been discussed here. Indeed, regular reference to the work of these individuals may be found throughout this book. Here we find critique of the prevailing focus on individual behavior change in much climate change education, and of post-political "win-win" framings of climate change education and action (Boström et al., 2018; Jorgenson et al., 2019). We find a repeated insistence that learning activities must "engage in issues of structural inertia, power, inequality, vested economic interests, denialism, resistance to change, and anxieties; that is, in the *conflict dimension* of sustainable development on societal and individual levels" (Boström et al., 2018, p. 2; see also Jorgenson et al., 2019). We find a concern with "places and geographies as transformative/transgressive forces" that should play a role in the development of effective and transformative climate change education (Lotz-Sisitka et al., 2015, p. 76; see also Selby & Kagawa, 2018). We find calls to link climate change education more closely with social movements and collective action (Jorgenson et al., 2019; Lotz-Sisitka et al., 2015). We even find the development of a "Living Spiral" as a model of transgressive climate change education, that has similarities with the popular education spiral discussed in Chap. 8 (although there is apparently no direct link between the two models) (Macintyre et al., 2020).

One of the striking conclusions that comes from reading this recent literature is that those who are currently working on developing transformative or transgressive approaches to climate change education are focusing on issues that have long been at the heart of radical social justice education: how to achieve fundamental social, cultural and economic change in the interests of the most marginalized, disempowered and oppressed groups in society. Yet, there is often a gap or disconnect between the field of radical education and this emergent literature on transformative climate change education. Take, for example, *The SAGE Handbook of Critical Pedagogies* (Steinberg & Down, 2020) that was published in 2020: in a collection that spans over one hundred chapters and one thousand five hundred pages, there are only a handful of mentions of the problem of climate change, and no attention paid to any of the work being done on developing subversive, transformative or transgressive approaches to climate change education. This is a missed opportunity, as there is clearly much that radical educators can learn from those engaged in this essential intellectual and pedagogical work. Climate change education is not only a vital site for doing radical or transformative education that can address the climate crisis: it is an increasingly important and useful site for learning how to do radical education than can address any number of other important issues as well. Intersectionality is at the heart of this work—for Heila Lotz-Sisitka and her co-authors (2016, p. 51), the "aspects of society … which require challenging for substantive sustainability transformations to emerge" include such concerns as "colonial practice or epistemology, gender and race relations, social exclusion, didactical contracts, environmental injustices" as well as other "structures of privilege, hegemonies of power" and arenas of "systemic dysfunction or systemic violence."

But there are also differences between the approaches in work on transformative and transgressive climate change education, and the approach taken in this book. One is the intellectual and political starting point. Much of this literature begins with the work of Jack Mezirow, who developed the concept of transformative learning to refer to "learning that transforms problematic frames of reference—sets of fixed assumptions and expectations (habits of mind, meaning perspectives, mindsets)—to make them more inclusive, discriminating, open, reflective, and emotionally able to change" (Mezirow, 2003; quoted in Aboytes & Barth, 2020, p. 994). Scholars and activists in this tradition note that "Mezirow's view of transformative learning is … [focused] mainly on cognitive transformation/s of individuals" and "does not fully theorise the relationship between cognitive transformations and social action or agency, especially collective transformation of human activity" (Lotz-Sisitka et al., 2015, p. 75). This gap is something that they then attempt to address in their own work. In other words, the field of transformative and transgressive climate change education tends to start with a model that is focused on individual and cognitive change, and seeks to add in concerns with collective, social and structural change (Boström et al., 2018). The starting point of this book with the older, broader traditions of radical education is different, for it begins with forms of education—popular education, social movement learning, community organizing, critical pedagogy, as well as many strands of democratic, feminist, and anti-racist education—that are already centrally concerned with collective learning and structural change in society. Much of this literature is either missing in recent discussions of transformative climate change education, or appears only parenthetically and sporadically. This is a conversational engagement that would be well worth developing, and hopefully the discussions in this book of these traditions of radical education in the context of the climate crisis can contribute toward this end.

Given these different points of origin, the diverse traditions of radical education have much to contribute to the project of developing a transformative or transgressive (or radical) approach to climate change education. For these traditions offer a set of theoretical concepts and educational practices for focusing on concerns of conflict, power and politics, and linking learning with projects of fundamental social transformation. In this book, this can be seen in everything from discussions of the hidden curriculum and pedagogy of hope through to models of democratic education, social movement learning, the cycle of community organizing and popular education spiral. The traditions of radical education provide us with a wealth of ideas for how to think specifically and concretely about the different kinds of things that need to be addressed in any model of education oriented to a project of radical social transformation: questions of power, positionality and place, politics and self interest, hope, fear and emotionality, and so forth.

The emergent literature on transformative climate change education tends to be positioned in reaction to dominant forms of climate change education that are focused on narrower concerns of building scientific literacy or effecting individual behavior change. K.C. Busch et al. (2019) compare transformative climate change education with positivist, post-positivist and social constructivist approaches; while Thomas Macintyre and his co-authors (2018) contrast "transformative social learning approaches" with science-oriented, policy-oriented and organisational and management-oriented learning approaches. One consequence of this framing is that the transformative climate change education literature tends to constantly foreground the need for sweeping and rapid transformation to address the challenge of the climate crisis. Transformative learning, according to Magnus Boström and his co-authors (2018, p. 7) "requires examining, questioning, and revising much of what we hitherto have taken for granted," it "includes changing frames of references and targets the meaning-making processes of individuals," and seeks to transform everyday "rules, norms, habits and practices." These repeated calls for transformation and transgression in this literature can come to feel exhausting. As Heila Lotz-Sisitka warns, "this is not a neoliberal chase after change" (Kulundu-Bolus et al., 2020, p. 116). In the traditions of radical education, the goal of fundamental social change tends to be presumed—this is, after all, a core part of how traditions of radical social justice education are defined in the first place. The literature is thus often able to focus more on the theoretical and practical details of the different ways in which such goals may be realized.

This brings up one of the most important contributions that the traditions of radical education offer to the project of developing transformative approaches to climate change education. For these approaches sometimes seem to suggest that everything and everyone must change completely to address the climate crisis effectively. Stefan Bengtsson (2019, p. 2) describes how some climate change education scholars and activists have moved from an interest in fostering transformative learning toward developing the concept of transgressive learning, because of their desire to "focus on discontinuity and breaking with continuity in learning," and pursue "a radical disengagement and disruption of existing systems of practices and norm systems." The danger of such an approach is that it risks coming across as yet another deficit model of blaming learners for creating the problems they confront: for utterly everything is wrong with us today.

Traditions of radical education take a different approach, as their project of using education to push for radical social transformation has long been balanced by and grounded in a clear commitment to begin by recognizing, appreciating and respecting where learners are in their lives currently: their places and positionality in the world; their capabilities, insights and voice, regardless of age or life stage; their self-interests, understood as being multiple and complex; their rights to speak, act and participate; their power as a community and a collective. It is this insistence on the value and strength of what already exists in all communities, and especially in marginalized and disempowered communities—that is so often ignored, denigrated or left unseen in many formal approaches to education—that is as central to the core nature of radical education as are its utopian projects of fundamental social change, empowerment and emancipation. In the Mississippi Freedom Schools of the US civil rights movement, for example, that sought to empower African-American students "to become a force for social change" to radically transform a fundamentally racist society, classroom discussions deliberately centered not just on what must change but also on what should be retained. "What do we have that we want to keep?," Freedom School learners were asked: "What does the majority culture have that we want? What does the majority culture have that we don't want?" (Chilcoat & Ligon, 1998, pp. 171, 175). The lesson is one that remains valuable for today's efforts to develop an effective climate change education. We don't only need to focus on what must change and be changed to be able to create a sustainable, just and healthy future for all, but on what we already have that we want to hang onto, protect and build upon, and what is in existence elsewhere around the world today that we want to learn from, adapt, and make our own.

References

Aboytes, J., & Barth, M. (2020). Transformative learning in the field of sustainability. *International Journal of Sustainability in Higher Education, 21*(5), 993–1013.

Balsiger, J., et al. (2017). Transformative learning and education for sustainable development. *GAIA, 26*(4), 357–359.

Bengtsson, S. (2019). Engaging with the beyond—Diffracting conceptions of t-learning. *Sustainability, 11*, 1–20.

Boström, M., et al. (2018). Conditions for transformative learning for sustainable development: A theoretical review and approach. *Sustainability, 10*(4479), 1–21.

Busch, K., Henderson, J., & Stevenson, K. (2019). Broadening epistemologies and methodologies in climate change education research. *Environmental Education Research, 25*(6), 955–971.

Chilcoat, G., & Ligon, J. (1998). 'We talk here. This is a school for talking': Participatory democracy from the classroom out into the community: How discussion was used in the Mississippi Freedom Schools. *Curriculum Inquiry, 28*(2), 165–193.

Jorgenson, S., Stephens, J., & White, B. (2019). Environmental education in transition: A critical review of recent research on climate change and energy education. *Journal of Environmental Education, 50*(3), 160–171.

Kulundu-Bolus, I., McGarry, D., & Lotz-Sisitka, H. (2020). Learning, living and leading into transgression. *Southern Africa Journal of Environmental Education, 36*, 111–130.

Kwauk, C., & Casey, O. (2021). *A new green learning agenda: Approaches to quality education for climate action*. Brookings.

Lotz-Sisitka, H., et al. (2015). Transformative, transgressive social learning: Rethinking higher education pedagogy in times of systemic global dysfunction. *Current Opinion in Environmental Sustainability, 16*, 73–80.

Lotz-Sisitka, H., et al. (2016). Co-designing research on transgressive learning in times of climate change. *Current Opinion in Environmental Sustainability, 20*, 50–55.

Macintyre, T. (2019). *The transgressive gardener*. PhD dissertation, Wageningen University.

Macintyre, T., et al. (2018). Towards transformative social learning on the path to 1.5 degrees. *Current Opinion in Environmental Sustainability, 31*, 80–87.

Macintyre, T., Tassone, V., & Wals, A. (2020). Capturing transgressive learning in communitites spiraling towards sustainability. *Sustainability, 12*, 1–20.

Mezirow, J. (2003). Transformative learning as discourse. *Journal of Transformative Education, 1*(1), 58–63.

Selby, D., & Kagawa, F. (2018). Teetering on the brink: Subversive and restorative learning in times of climate turmoil and disaster. *Journal of Transformative Education, 16*(4), 302–322.

Steinberg, S., & Down, B. (Eds.). (2020). *The SAGE handbook of critical pedagogies*. SAGE.

T-learning. (2021). About. https://transgressivelearning.org/about/

Index

S. Tannock, *Educating for Radical Social Transformation in the Climate Crisis*, Palgrave Studies in Education and the Environment,
https://doi.org/10.1007/978-3-030-83000-7

The manufacturer's authorised representative in the EU is Springer Nature Customer Service Centre GmbH, Europaplatz 3, 69115 Heidelberg, Germany. If you have any concerns regarding our products, please contact ProductSafety@springernature.com

Printed and bound by CPI Group (UK) Ltd, Croydon, CR0 4YY

30/09/2026

02225702-0011